Fish, Markets, and Fishermen

Fish, Markets, and Fishermen

The Economics of Overfishing

Suzanne Iudicello,
Michael Weber,
and
Robert Wieland

Center for Marine Conservation

Earthscan Publications Ltd, London

First published in the UK in 1999 by Earthscan Publications Ltd
First published in the United States in 1999 by Island Press

A catalogue record for this book is available from the British Library

ISBN: 1 85383 651 6 (pb)
1 85383 652 4 (hb)

Typesetting by Brighid Harpe
Page design by Island Press staff
Printed and bound by Edwards Brothers
Lillington, North Carolina, USA

For a full list of publications, please contact:

Earthscan Publications Ltd
120 Pentonville Road
London N1 9JN, UK
Tel: +44 (0)171 278 0433
Fax: +44 (0)171 278 1142
email: earthinfo@earthscan.co.uk
http://www.earthscan.co.uk

Earthscan is an editorially independent subsidiary of Kogan Page Ltd and publishes in association with WWF-UK and the International Institute for Environment and Development.

Contents

Foreword

We have overfished marine fish populations for hundreds of years. Why do we continue? Are we greedy? Weak? Are we blinded by the romance of freedom, adventure, and courage that fishing the ocean represents?

The reason that overfishing continues is not because people are greedy, weak, or overly romantic but because it is a rational thing to do.

When a fishery is open to anyone, there is no assurance that a fish not caught today will be around tomorrow. In fact, it will probably be caught by someone else. So why not catch it yourself? Why invest in the long-term sustainability of the fishery if what happens tomorrow or next week or next year is highly uncertain? It's not rational. The only rational thing to do is to race for the fish, to fish early and often, and to build a boat that will out-fish competitors.

That is the incentive of open access. The uncertainty among fishermen about whether fish will be there tomorrow interacts with the economics of fishing and the natural variability of fish populations in a way that is destructive of the resource. Attempts by the government to manage fishing behavior are inadequate to control these dynamics.

Leaving fisheries open to anyone is often done with the best of intentions. It is an attempt to be fair to all and to share the opportunity of fishing. But despite this noble goal, the outcome of open access fishing is neither fair nor profitable. As this book clearly demonstrates, an open access fishery will attract too many boats, too many people, and too much fishing power. The result is an overfished fishery.

This book is first and foremost about fisheries as human systems, through which biology, economics, and politics interact. It is about the sad history of open access fisheries and their inevitable outcome. It is about how, in attempting to help the fishermen with subsidies and assistance, the government has, in the end, hurt them by allowing them to expand too much. And it is about understanding that it is not through greed or weakness that open access fisheries are overfished, but rather through uncertainty and an inability to control the resource on which the fisherman's livelihood depends.

And, perhaps most importantly, this book is about how an understanding of the economics of fishing can be used to protect fishery resources by turning around the incentives to overfish.

This book shows in clear nontechnical language how biology, economics, and politics are forever linked in fisheries, and how knowing about these linkages will help us understand the problems of fisheries and craft solutions to these problems. Through theory and case studies, the authors show how economics drives fishing behavior, how subsidies hurt, how management tools work, and how economic tools can contribute to solutions. Through a legislative history of the 1996 reauthorization of the Magnuson Fishery Conservation and Management Act (now the Magnuson-Stevens Fishery Conservation and Management Act), the authors illustrate the political process of crafting fishery policy.

Case studies from the United States, Canada, and New Zealand present details of how economic tools work in different types of fisheries. Called individual fishing quotas (IFQs), they are a "right" to catch a certain share of the quota. A quota share is assigned to an individual fisherman. Once he has an IFQ, a fisherman doesn't have to race for his catch. He doesn't have to build a bigger boat to catch it all at once. He can decide when it makes the best sense for his business to catch and sell the fish. He can sell his quota shares, lease them for a season, or buy more if he wants to expand his business.

Because IFQs are so different from the regulations we have traditionally used, and because they move the competition of fishing away from hunting and into business, they are controversial. They are so controversial that Congress placed a moratorium on their use in 1996, and required that a National Research Council (NRC) study be performed of their properties and effects before they could be used in fishery management.

The NRC Committee on the Use of Individual Quotas in Fisheries has now completed its examination and issued its report.[1] In the course

1. National Research Council. 1999. *Sharing the Fish: Toward a National Policy on the Use of Individual Quotas in Fisheries.* Washington, D.C.: National Academy Press.

of its work the NRC met in five fishing regions and listened to many hours of testimony about the experience with individual fishing quotas. It found, consistent with the case studies presented in this book, that like any fishery management tool, IFQs have both costs and benefits.

The committee concluded that the performance of an IFQ system depends on how it is designed to fit the context of a particular fishery. Concerns about the fairness of quota-share allocation, concentration of quota ownership, and the position of traditional fishing communities must be addressed. The committee recommended that although IFQs might not be appropriate for all fisheries, they are an effective tool for limiting access that should be available to regional fishery managers: The moratorium on using these economic tools of management should be lifted.

The controversy over the use of individual quotas demonstrates a truism of fishery management that is the central lesson of this book. Fishery management is never just about fish and never just about people. It is always about the complicated mix of fish, economics, cultural traditions, politics, and law. It is about the ideals that we have for marine resources, the goals we form, and the choices we make as these ideals become modified by practice.

American fishery resources are owned by the people and are managed with participation of the people. Management relies on many different types of fishery stakeholders: industry, environmental organizations, government, academics, consumers, and others who care about fish. The success of this system rests on the quality of that participation, which can only be improved by absorbing the lessons of this book.

Read it, understand the incentives shaping fishing behavior, and be ready to contribute to the debate about how to turn these incentives around to protect the future of fisheries.

Susan S. Hanna
Professor of Marine Economics
Oregon State University

Preface

When the first edition of this book was published, in 1992, the economics of fishing and fishery management were not generally part of the conversation around the coffee urn at fishery management council meetings. Topics such as license limitations and quota shares, though familiar in a few isolated fisheries, had barely bubbled up to awareness among fishing folk, let alone become issues in the news media or with the general public.

In contrast, as we go to press in 1999, major newspapers in the United States have reported on the findings of a controversial study by the National Research Council indicating that limited access to fisheries and various options for managing with quota shares will be necessary tools for fishery managers in the coming years.

Economics may have been dubbed the "dismal science" in the past, but in the fisheries context, the issues are far from dismal. The national policy debate over whether to limit access in U.S. fisheries and how to structure quota programs in individual fisheries is not just heated; it is boiling over.

With an eye to the next revision of the U.S. law that governs fishery management, we hope to inform the debate about whether managers should be allowed to make use of economics-based tools such as quota programs and whether to end the open access regime that is threatening fishery resources.

We hope to fill a need for an understandable book that is not only about economic theory and not only about fishery management but links the two with real-world case studies, examples from the fishing industry as it operates, and illustrations that explain the theory with examples of individual behavior in actual fishing situations. With this book, we aim to tell the story of how skippers on fishing vessels make decisions during fishing operations and how those decisions differ under open access, limited access, and various property rights systems.

For those seeking a definitive text on fishery economics, we offer references and further reading. This volume does not presume to be such a work. Rather, it is written for those most affected by the workings of economic forces that underlie fishery operations and that are driving fish populations to dangerously low levels. Although this audience of stakeholders may not particularly care about the elaborate algebraic equations that illustrate the theory, an improved understanding of the economics may contribute to participation in fishery management and inform the debate about management options.

It is to those stakeholders—fishers, fishery managers, politicians, fish advocates, and the fish-consuming public—that we direct and dedicate this book.

The authors thank members of the Center for Marine Conservation, whose generous support made this work possible. We also acknowledge the countless fishery managers, biologists, and economists who answered questions and reviewed drafts. Suzanne Iudicello wishes to thank especially Lee Anderson for his painstaking review and Susan Hanna for her support and encouragement. Michael Weber owes an enormous debt of gratitude to Jon Goldstein for patiently enticing and then introducing him to the world of economics.

Introduction

Gloucester, Massachusetts, 1992

For three generations, a fishing family relied on the cod, haddock, and flounder off New England's coast. In the grandfather's day, jumbo codfish upward of 50 pounds were not uncommon. The father recalls pulling up 5,000 pounds of cod in a one-hour tow. Today, the son "might haul in 2,000 pounds of middling-sized cod in eight hours of hard trawling."

"The Rape of the Oceans,"
U.S. News & World Report, June 22, 1992

Astoria, Oregon, 1992

A former salmon fisher has fallen back on sewing nets, hoping that government disaster relief will provide her with money for education. She lost her boat, her house, and the fishing business she had hoped her son could continue. "I was doing really well in the eighties, but then my income kept dropping, until this spring I couldn't afford the $385 fishing license. And you can't renew the license if you don't deliver fish on it."

Carl Safina, *Song for the Blue Ocean,* 1998

Newfoundland, Canada, 1992

More than 20,000 fishermen and fish processors were abruptly put out of work when the government shut down the Grand Banks, "where once-vast schools of cod, numbering in the hundreds of millions, had dwindled away."

Deborah Cramer, "Troubled Waters," *Atlantic,* June 1995

Gulf of Alaska, 1994

Over the years, effort in the halibut longline fishery had increased so much that the fleet was ratcheted down to two short openings per year. In September 1994, halibut long-liners desperate to make a catch in the two-day opening put to sea despite gale force winds and a storm that covered the entire coast of Alaska. One life and six vessels were lost in that last derby opening in the halibut fishery.

Cape May, New Jersey, 1994

Where fishermen once could fish for a variety of species without much regulation, they are now restricted to a few remaining stocks as rules tighten to protect the growing number of dwindling species. "Last summer the *Adrianna* trawled for scallops. And then for groundfish, such as flounder. But last month new federal rules severely restricted scalloping in the Atlantic, and after an eight-week season, the schools of winter flounder were closed completely. So it's squid."

"Overfishing, New Regulations Curtail Livelihoods on the Sea,"
Philadelphia Inquirer, April 6, 1994

Newfoundland, Canada, 1994

"In April, Canadian officials charged into international waters off Newfoundland and impounded the Portuguese-owned trawler *Kristina Logos*. . . . Canada has banned catching cod and most other groundfish within its own 200-mile limit, hoping to save its once-rich Grand Banks fishing ground. The April seizure expanded that effort and Brian Tobin, Canada's Fisheries & Oceans Minister, promises more vigilance."

Business Week, July 4, 1994

Gloucester, Massachusetts, 1994

That same month, "a procession of fishing boats staged a demonstration in Boston Harbor, and a group of outraged workers protested in Gloucester, turning over cars and dumping fish off a truck. Massachusetts Governor William Weld promised $10 million in aid to fishing towns and called for federal help. That came last week, when the Clinton Administration announced a $30 million aid package for New England."

Time, April 4, 1994

Nova Scotia, Canada, 1995

"All along the coast of Nova Scotia boats lie idle at the dock."

Deborah Cramer, "Troubled Waters," *Atlantic,* June 1995

In March, Canadian fishery protection officers seized the Spanish fishing vessel *Estai* for violating fishing rules for Greenland turbot, pushing the limits of international fishing law.

Talumphuk, Thailand, 1996

"For generations, the men of Talumphuk have left their thatched huts each morning, heaved their boats down log runners, and skimmed off to set their nets for the valuable shrimp and bottom fish that teemed nearby in the Gulf of Thailand. . . . 'Ten years ago, we could catch anything we wanted,' said Sophon Losere-sakun, wearing a red loincloth and crouched on the bamboo deck of his house. 'But in the last 10 years, it has gotten to be less and less. Now we have almost nothing.'"

John McQuaid, "Overfished Waters Running on Empty," *New Orleans Times-Picayune,* March 24, 1996

Standing on the shore and looking out to sea, people have thought that the oceans could always be a source of food and natural resources. Trends in abundance of living marine resources in the latter part of the twentieth century, however, have provided compelling evidence that we have reached and in some cases exceeded the productive limits of the ocean. The status of ocean fisheries indicates that we do in fact have the means to deplete some populations of fish.

Why is this happening? Who's to blame? Why haven't we prevented this result?

Health-conscious consumers who want to eat more fish, fishermen who are trying to meet that demand and make a living, divers who admire colorful reef species, anglers who enjoy the solitude or the challenge of fishing, and environmental advocates who want to protect fish as an integral part of a healthy ocean ecosystem could all agree that no one benefits without enough fish in the sea.

Yet world production of fish has been on a flat or downward trend since the late 1980s. The Food and Agriculture Organization of the United Nations once estimated that the potential catch of traditional fisheries (excluding aquaculture) was 100 million tons per year. Now, however, that agency has concluded that we cannot reach that potential level of an important food source without drastic change in the way we manage fisheries. Whether one is a fisher, a fish consumer, or just a person who likes the thought of healthy ecosystems, the decline of many of the world's fish populations is of concern.

With increasing interest in ocean issues and greater public awareness of threats to marine resources, the rhetoric about who is to blame for the increasing scarcity of marine fish is escalating. Is it fishermen who prefer taking their income now rather than conserving for the future? Is it the recreational sector, with its sonar and fast boats? Is it pollution from land-based activities? A flawed management system? Lack of understanding of the biology of fish populations? Political intervention?

All these factors contribute, to be sure. This book aims to describe in simple terms the economic and biological reasons why overfishing could have been predicted, given current technology and consumer demand for fish. To do this, we describe how fish populations, human consumers, and modern-day technology all interact to the detriment of the fish. Under this model, blame can be ascribed to everyone pretty much equally.

In this book, we explore a set of relationships among fishermen and the biological, economic, and technological factors that affect the fisheries they pursue. Through both theory and example, we demonstrate what happens in a system in which access to newly discovered fish resources is open to all comers. The early period is characterized by high catches, growth, and profits for the first to enter the fishery. This profit bonanza then attracts more newcomers or more vessels. As time goes on, participants have to fish harder to increase or even maintain their catches by using more or more efficient gear, increasing the size of their boats, or fishing longer. Because the amount of fish that can be caught has its own biological limits, the amount caught by each boat will decline. To compensate for the higher cost of catching the same amount of fish, fishermen add technology to increase their fishing power, time, and effort. The cycle of more effort being expended in chasing fewer fish continues, increasing costs for participants to the point that some fishermen may have to leave the fishery. This progression is described in figure I.1. Each aspect of this diagram is discussed in detail in the chapters that follow.

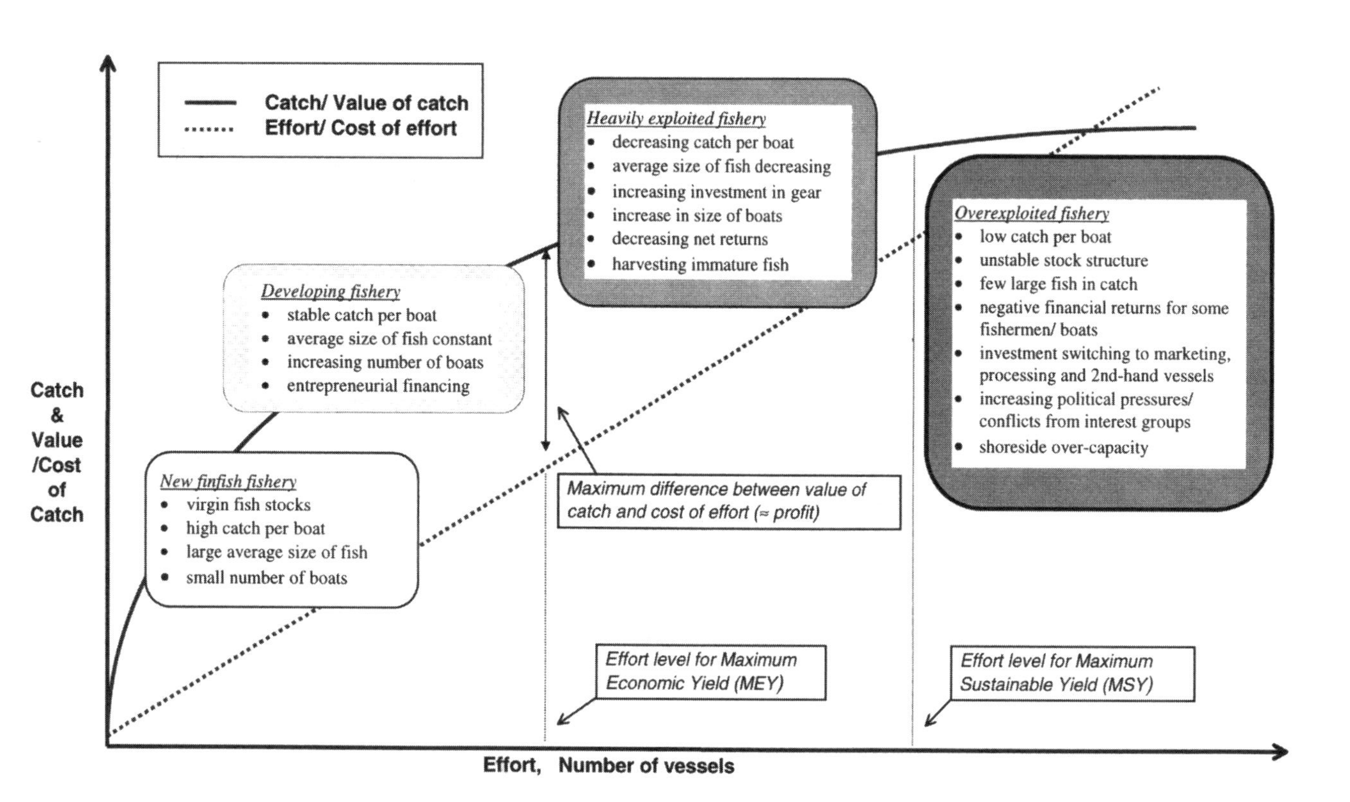

Figure I.1. Growth of an uncontrolled open access fishery (*source:* M.K. Kelleher. Approaches to Practical Fisheries Management. FAO Roundtable on management and regulation of fisheries, Dakar, Senegal, 1–3 July 1996.)

In chapter 1, we show that some marine fisheries are overfished. A description of world fish populations and the fleets that pursue them portrays the declining trend in catches despite increasing trends in vessels and fishing power. This topic should be of concern to fishers because it points to a greater potential for commercial failure than would be the case if stocks were larger. It should be of concern to fish consumers because fewer fish probably translates into more expensive fish. And it should be of concern to those interested in healthy ecosystems because it means that an important component of the marine ecosystem is missing.

We then discuss the economics of fishing and why, given open access and advanced technology, overfishing and diminished fish stocks should be expected. This economic view provides a useful way to address what is happening in the marine fishing industry because it illuminates the incentives to which fishers and consumers respond in creating overfishing. It also allows us to describe more effective ways for fishery managers to bring about a return to healthy stocks that are not overfished.

Economics: The Study of How Scarce Resources Are Allocated in Markets

Economics is the study of how limited resources are allocated among unlimited wants. We all want things—call them goods and services—and we acquire them to the degree we can afford to do so. Economics is the study of how this happens, primarily in markets but also in households and institutions. In this book, we focus largely on the economics of markets.

One can tell a great deal about the flow of resources by studying the markets in which they are bought and sold. Markets are "virtual places" where demanders and suppliers hammer out not only the prices at which resources will change hands but also the amount of a resource that will be supplied. In chapters 2 and 3, we describe how in market economies consumers (demanders) draw forth the supply of goods and services through their willingness to pay and how both producers and consumers gain. In chapter 2, we show how goods are supplied to consumers when all factors of production are owned. This story is based on the economic theory of the firm, the economic explanation of supplier behavior, which is a powerful descriptive tool in microeconomics. Then, using the example of fishers working in the open sea, we show how things change when supply of the goods involves an unowned but limited renewable resource.

In chapter 3, we describe important aspects of the natural production of fish and how these affect the supply of fish to markets. We also describe the incentives that help determine the level of effort applied to catching fish. In the description of markets for fish and the natural production of fish, we show why it should be expected that if they can, suppliers will apply too much effort to catching fish and in the process will catch more than the fish population can replace through natural production.

For many years, fishery managers set catch limits and fishing rules based on their understanding of the response of fish populations to mortality from fishing. The fundamental management concept—maximum sustainable yield—was premised on the notion that people will remove fish from the ocean, and the goal of management was to enable people to remove as much as possible. Maximum sustainable yield is the largest annual catch fishers can take from a stock over the long term without changing its ability to continue to produce that surplus under existing environmental conditions. The emphasis was on the yield of single species for human use. Very often, the emphasis on maximizing yield has subverted even the best intentions. Fishermen who might otherwise have left a fishery that was no longer profitable were lured back by subsidies that enabled them to increase fishing power with low-cost government loans. For example, government subsidies of fishing, the main subject of chapter 4, have been widely used to build capacity in the industry.

And it is no wonder that government support of fishing is so common: people use wild populations of fish and shellfish directly as food and to produce oil, fish meal for animal food, and fertilizer for crops. Worldwide, 15–20 million people fish for a living, and 90 percent of them are small-scale fishers. Fisheries provide as many as 180 million additional jobs in associated sectors such as processing, packing, and distribution. However, recent trends in catch, trade, contribution to food supplies, and overall economic viability of fisheries are declining.

These trends, captured in the headlines of the mid- to late 1990s, have brought the plight of ocean fish to public awareness not only in coastal areas but also in landlocked regions. This public sentiment in favor of conservation has made a difference in government response to declining fisheries. In chapter 5, we discuss how fish as a natural resource have come under public stewardship and how fishery management has developed in that realm. We also describe government response to overfishing and the legal framework for fishery management.

We then look at some of the tools fishery managers use to control fishing effort, applying concepts introduced in the previous two chapters to show which outcomes should be expected when managers try to control fish catches by means of one or another of these tools. Economic analysis generally makes it possible to compare the effectiveness of these tools in achieving management objectives.

Chapter 6 presents a series of case studies that show how these concepts have played out when applied to actual fisheries. From the small-boat salmon fisheries of Alaska to the orange roughy industrial fleet of New Zealand, from fisheries as varied as Atlantic surf clams and Pacific halibut, the case studies examine attempts to devise management programs that incorporate economic principles, exploring the level of success these programs have achieved in limiting effort and stopping the race to fish.

Finally, in chapter 7 we take a look at the most recent government response to rights-based approaches in the United States and recent revisions to the Magnuson-Stevens Fishery Conservation and Management Act. We conclude that U.S. lawmakers have—at least temporarily—deprived managers of an important tool for fishery conservation.

How Much Is Too Much?

Those who manage fisheries generally agree that among the pressures on these resources, the two critical ones are overfishing and overcapacity. In this book, we conclude that the two are inextricably linked in systems that are open to access.

Overfishing is the taking of so many fish from a population that the stock's capacity to produce maximum sustainable yield on a continuing basis is diminished. Some fishery biologists argue that even maintaining maximum sustainable yield is not sufficiently conservative to protect populations from overfishing. The preference for maximum sustainable yield as a target has yielded to the idea that it may be a ceiling. Consideration of natural mortality, growth, and maturation rates and of annual recruitment of young into the population must figure into the equation. In addition, fishery biologists now urge consideration of fish as part of the marine ecosystem rather than as single species units in isolation. Fishing not only results in direct removal of the target animals but also affects the composition, abundance, and population structure of the target species and other species as well as the overall structure and function of the ecosystem. This is important because most fisheries consist of mixed stocks—that is, groups of species that are caught together but may not have the same abundance, growth rates, or life histories.

The favoring of resource use in the absence of information about the effects of such use has given way to the precautionary approach, which demands that in the face of uncertainty, managers must err in favor of protecting living resources. This shifts the burden of proof (that a technology or level of effort is not harmful) to the user of natural resources. The precautionary approach has become part of international agreements and is integrated into the national management strategies of many nations. Such shifts in policy and practice were unheard of until recently, partly because marine fish conservation was a somewhat difficult issue to champion. Unless they were presented tastefully on a plate, fish were out of sight, out of mind.

Overcapacity (or *overcapitalization*) is an excessive level of catching power, more effort in terms of vessels, time, and gear than is necessary to catch the amount of fish available. It is this element of the two critical pressure points that continues to drive overfishing, no matter how precautionary are the biological approaches chosen by managers. Indeed, most of the management measures that have been tried, and that are explored in detail in this book, have proven incapable of conserving fishery resources. The economic theory laid out on the following pages explains why so much effort is brought into fishing and why, under open access, that will continue.

✦ Chapter 1 ✦

Not Enough Fish in the Sea

Until recently, in the balance between productivity of fish populations and people's ability to catch fish, the fish were favored.[1] Although fishing did deplete some fish populations before World War II, most of them survived the rudimentary technology of the times. Cotton nets, hand lines, coastal vessels with short ranges, and the individual fisherman's eyesight, experience, and fish-finding capability circumscribed fishing capacity. This balance changed rapidly, however, as innovations triggered by research and development for the war effort were introduced into fishing fleets. Fiberglass for lighter and cheaper hulls, synthetic line for larger and lighter nets, diesel engines, and electronic gear for locating productive fishing grounds suddenly increased fishermen's ability to find and catch fish.

Innovations in processing, transport, and marketing of fish increased its availability in some countries. These innovations, together with rising human populations and affluence in western Europe, the United States, and Japan especially, increased demand. Whether motivated by dreams of riches or dreams of feeding the world's poor with cheap protein, governments and businesses invested in more and larger fishing vessels and infrastructure to tap seemingly inexhaustible fish populations.

Now, in many fisheries around the world, fishing fleets are far larger than is necessary to catch the amount of fish that fish populations can produce over the long run. This imbalance has already caused economic and ecological dislocation in some of the world's premier fisheries, such as the groundfish fishery off New England and Newfoundland.

Trends in World Fisheries

Far from being a local affair, fishing is a global business that generates billions of dollars in trade and fuels economic development in several countries. Since World War II, landings of marine fish and shellfish have grown worldwide, from 16.3 million metric tons[2] in 1950 to 91.3 million metric tons in 1995 (FISHSTAT 1997) (see figure 1.1). Between 1950 and 1976, when the United States and many other countries extended their fishery jurisdiction to 200 nautical miles[3] off-

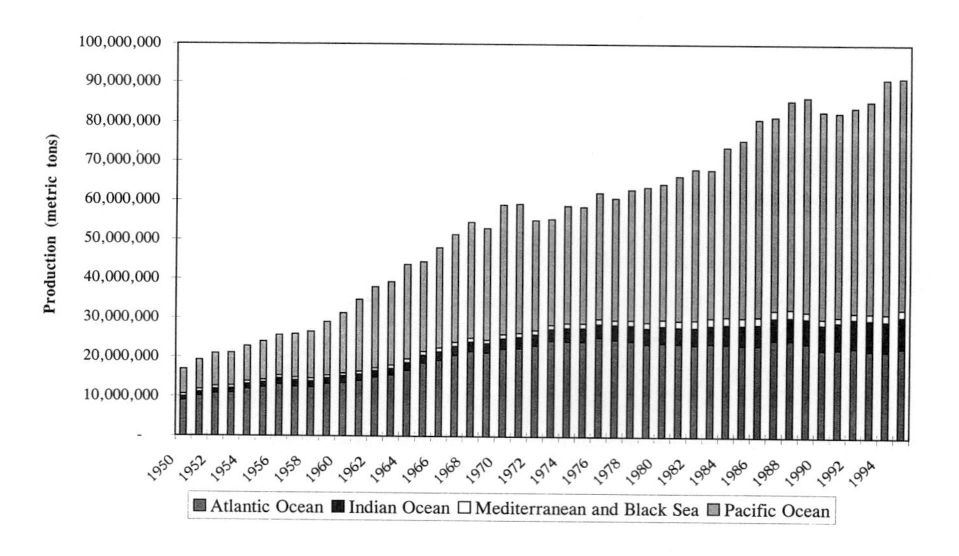

Figure 1.1. Worldwide Production of Marine Fish and Shellfish (*source:* FISHSTAT 1997)

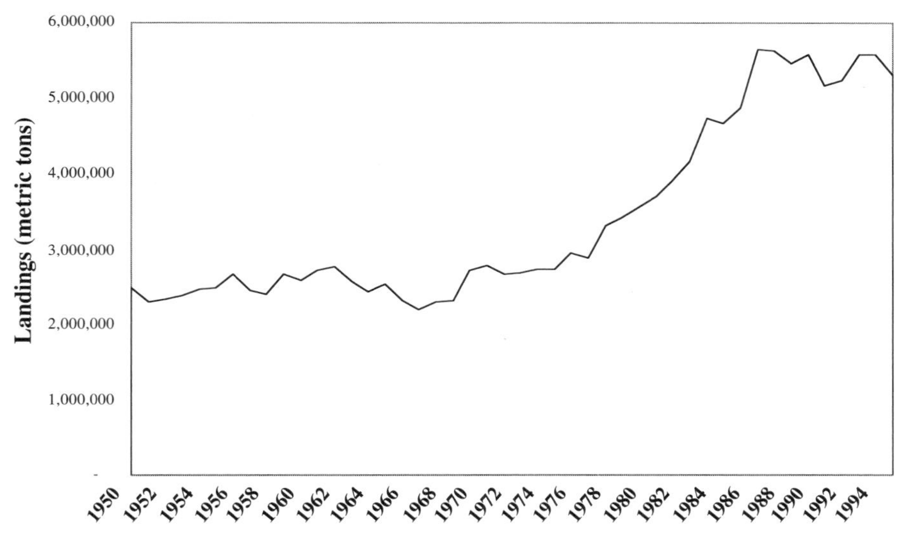

Figure 1.2. Total U.S. Marine Landings, 1950–1995 (*source:* FISHSTAT 1997)

shore, landings grew at an annual average rate of 5 percent (see figure 1.2). The rate of growth in landings declined to 1.7 percent in the 1970s and then grew to 3.6 percent in the 1980s (Garcia and Newton 1997). Landings peaked at 86.4 million metric tons in 1989 and then fell for only the third time since 1950 (FISHSTAT 1997).

Since 1950, pelagic (open ocean) species such as sardines, tuna, and mackerels have dominated world landings, accounting for more than 60 percent of the world catch in 1994 (Grainger and Garcia 1996). The importance of pelagic species in landings varies among regions. For instance, in the Pacific Ocean, pelagic species have accounted for as much as 59 percent of the catch, whereas in the Indian Ocean, they account for less than half. The decline in the Peruvian anchoveta in 1970 largely explains the slowing in the rate of growth in landings during this period.

Many pelagic fish populations undergo large, decade-long shifts in abundance over entire ocean basins. These fluctuations are thought to be climate-driven and account for the ups and downs in overall landings. For example, a dramatic decline in populations of Peruvian anchoveta in the early 1970s largely explains the decline in the rate of growth in landings during that period (Garcia and Newton 1997).

Similarly, a decline in South American and Japanese pilchard accounts for much of the decline in worldwide landings in the early 1990s.

The other main group of fishes are demersal fishes, that is, species associated with the ocean bottom. This group, which includes most of the higher-value fishes, accounted for half the value of world landings in 1993, compared with 40 percent for pelagic species (Grainger and Garcia 1996). A decline in demersal species such as cod and haddock in the northwestern Atlantic Ocean contributed to a slower growth rate in world landings in the 1980s (Garcia and Newton 1997). From 1970 to 1992, landings of the four principal demersal species—Atlantic cod, silver hake, haddock, and Cape hake—declined by 67 percent, from 5.0 million metric tons to 1.6 million metric tons.

This period also saw a dramatic increase in international trade in fish products. Between 1950 and 1980, the percentage of fish landings that entered international trade rose from 20 percent to 33 percent (OECD 1997). Between 1980 and 1993, the value of traded fish products rose from $2.9 billion to nearly $40 billion (OECD 1997). The deficit in trade among developed countries, which accounted for 85 percent of imports by value, grew from $700 million in 1969 to about $15 billion in 1990 (Garcia and Newton 1997). Of the developed countries, only the United States decreased imports, entirely as a result of its declaration of a 200-nautical-mile fishery zone off its shores, which includes the enormous Alaska groundfish fishery.

Developing countries increased their share of exports from 32 percent in 1969 to 44 percent in 1990 while keeping their imports below 13 percent of world trade (Garcia and Newton 1997). As a result, these countries increased their positive trade balance from $500 million to $10.6 billion. In many developing countries, high-value species such as shrimp and tuna are exported and lower-value species enter local and national markets.

Roughly 60 percent of world fish landings are destined for direct human consumption; the balance is processed, principally into fish meal and fish oil (FISHSTAT 1997; OECD 1997). Between 1950 and 1982, the percentage of catch distributed fresh fell from nearly half to 20 percent (OECD 1997). Improvements in freezing techniques led to slightly more than a fourfold increase in the percentage of fish that entered markets frozen, from 5 percent to 22 percent.

Real prices of fish rose after 1970, fell slightly in the early 1980s, and then rose again slightly (OECD 1997). By 1994, the average price per ton ranged from $235 for herring, sardines, and anchovies to $11,800 for lobster. As a group, the price of cod, hake, and haddock rose from $700 per ton in 1989 to $1,060 per ton in 1994, reflecting declines in abundance of these groundfishes. Prices of other species declined. Most dramatically, the price of salmon and trout fell from $3,500 per ton to $2,750 per ton between 1989 and 1994. An increase in salmon farming explains most of this price decline. By 1995, salmon farms in Norway, Chile, Scotland, and other countries were producing about one-third of the salmon entering the market.

Leading Fishing Countries

Several factors contributed to a shuffling of the leading fishing countries in the late 1980s. The fall of the Soviet Union and the consequent loss of state subsidies caused many ocean-roving vessels of the former Soviet Union to tie up at docks. Rising labor costs, among other things, caused Japan's government and fishing industry to move from depending on domestic fleets for supplies to using fleets from other countries, particularly Taiwan and Korea (Weber 1997a). Recovery of the Peruvian anchoveta renewed Peru's fisheries.

The need of developing countries to generate foreign exchange also contributed to changes in the top fishing countries. Of the top ten fishing countries in 1995, six—China, Peru, Chile, India, Indonesia, and Thailand—were developing countries. One other country, the Russian Federation, was classified as an economy in transition. Japan and the United States ranked fourth and fifth, after China, Peru, and Chile.

With the exception of Peru, developing countries have shown dramatic and steady increases in landings of marine fish and shellfish. Between 1977 and 1995, Chile increased its landings from 1.3 million metric tons to 7.5 million metric tons, and India increased its landings from 1.4 million metric tons to 2.7 million metric tons. During the same period, marine production by Indonesia grew from 1.2 million metric tons to 3.3 million metric tons and Thai production grew from 2.0 million metric tons to 3.1 million metric tons (FISHSTAT 1998).

In 1995, China dominated world production of marine fish and shellfish, accounting for 13.5 million metric tons, or nearly 15 percent of the world total (see table 1.1). By comparison, China ranked sixth in 1986, producing just 4.5 million metric tons. China's meteoric increase in fish production reflects a government program to increase production to 20 million metric tons by the year 2000 (Milazzo 1997). This ambitious program is aimed at securing food for China's growing population and generating foreign exchange. Between 1986 and 1995, China increased its exports of fish products from less than $500 million to nearly $2.9 billion (see figure 1.3).

In the late 1980s and the 1990s, Peru's landings rose rapidly as the anchoveta fishery recovered from record lows associated with overfishing and an intense El Niño in the early 1970s and early 1980s. In 1995, Peruvian fishers landed 8.9 million metric tons of fish, up from a low of 1.5 million metric tons in 1984. Peru's exports of fish products, largely fish meal, grew from $258 million in 1986 to $870 million in 1995 (see figure 1.4).

Chilean landings have risen steadily since the early 1970s, reaching a peak of 7.8 million metric tons in 1994. In 1995, Chile ranked third worldwide, with landings of 7.5 million metric tons. Chile's landings, like those of Peru, are dominated by small pelagic species, including

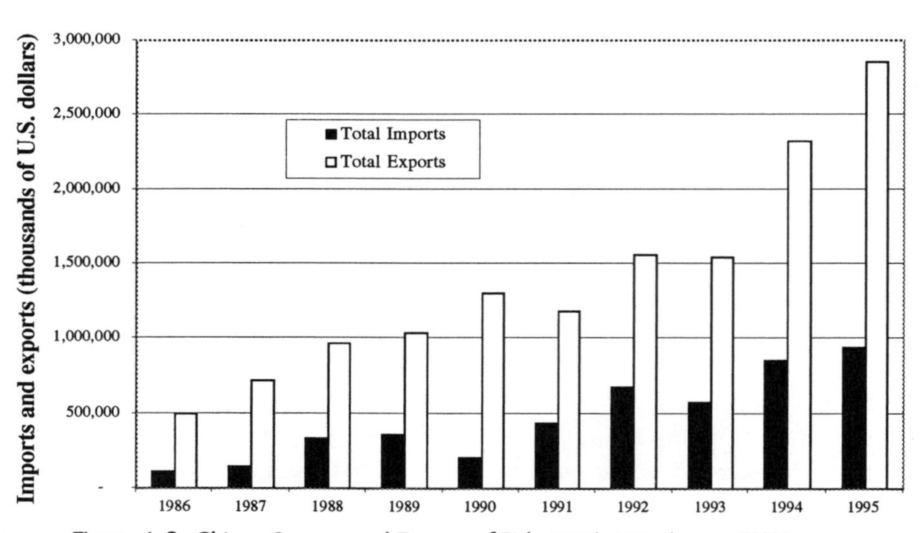

Figure 1.3. Chinese Imports and Exports of Fish, 1986–1995 (*source:* FISHCOMM 1997)

Table 1.1. Landings of Marine Fish and Shellfish by Leading Countries

Country	*1986*	*1987*	*1988*	*1989*	*1990*	*1991*	*1992*	*1993*	*1994*	*1995*
China	4,498,717	5,216,038	5,774,636	6,327,847	6,818,538	7,500,802	8,561,756	9,933,822	11,556,241	13,480,766
Peru	5,583,023	4,549,901	6,601,067	6,818,643	6,843,580	6,858,145	7,473,245	8,969,342	11,954,819	8,891,565
Chile	5,540,717	4,784,594	5,181,879	6,423,306	5,171,951	5,973,684	6,466,731	5,997,883	7,793,628	7,530,305
Japan	11,657,812	11,499,317	11,680,085	10,859,917	10,045,495	8,995,461	8,163,090	7,798,132	7,105,966	6,465,107
United States	4,872,118	5,639,652	5,632,325	5,450,945	5,567,900	5,165,021	5,243,571	5,569,236	5,582,455	5,294,835
India	1,730,944	1,693,739	1,806,637	2,259,124	2,220,193	2,388,667	2,510,815	2,615,110	2,733,263	2,699,550
Russian Federation	—	—	7,745,807	7,773,768	7,186,487	6,596,565	5,225,833	4,147,941	3,494,786	4,094,516
Indonesia	1,878,907	1,980,577	2,146,088	2,271,828	2,354,239	2,674,673	2,721,199	2,885,684	3,107,759	3,287,080
Thailand	2,271,991	2,561,283	2,432,890	2,476,892	2,542,842	2,655,574	2,866,092	3,037,783	3,073,404	3,134,200
Norway	1,914,751	1,949,026	1,839,613	1,907,881	1,746,217	2,172,274	2,560,901	2,561,266	2,550,750	2,807,136
SUBTOTAL	39,948,980	39,874,127	50,841,027	52,570,151	50,497,442	50,980,866	51,793,233	53,516,199	58,953,071	57,685,060
WORLD	80,904,566	81,558,662	85,605,656	86,438,889	82,925,830	82,614,855	83,922,644	85,483,931	90,987,509	91,322,931

Source: FISHSTAT 1997.

Note: All landings are in metric tons.

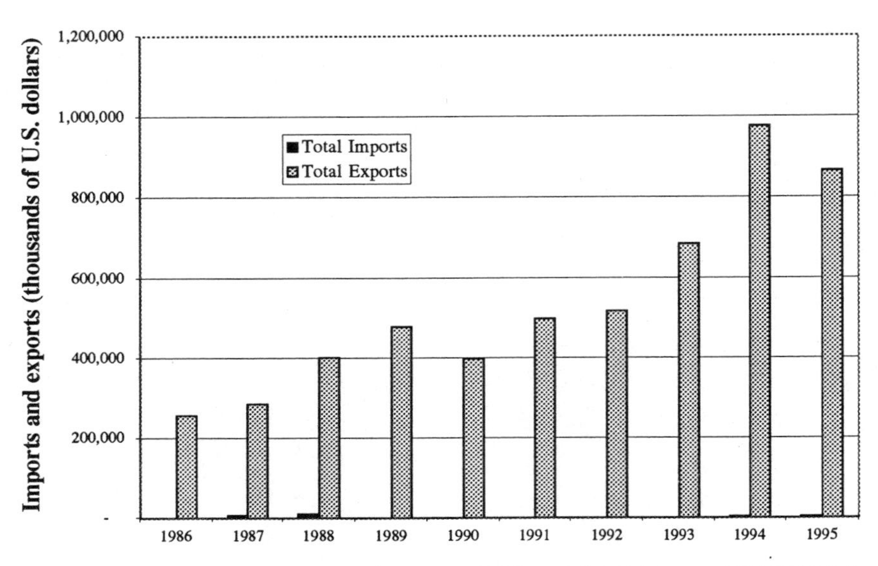

Figure 1.4. Peruvian Imports and Exports of Fish, 1986–1995 (*source:* FISHCOMM 1997)

mackerel and anchovies. Both Chile and Peru are relatively minor consumers of fish and export most of what they catch. In Chile's case, exports have grown in recent years, particularly as a result of production of Atlantic and Pacific salmon on farms; salmon farming grew from about 4,000 metric tons in 1986 to more than 150,000 metric tons in 1995 (AQUACULT 1997). During the same period, the value of Chile's exports grew from $515 million to $1.7 billion (see figure 1.5).

In the mid-1980s, when Japanese landings peaked at 11.7 million metric tons, coastal countries from the United States to the small islands of the Pacific began closing their waters to foreign fishing vessels. What had been a free resource for Japan's wide-ranging fishing fleets began carrying a cost imposed as access fees (Weber 1997a). Japanese processors began obtaining fish from other countries, especially Taiwan and Korea. Japanese landings began a steady decline in 1989, reaching a low of 6.5 million metric tons in 1995. During the same period, the value of Japanese imports rose from $10.1 billion to $17.8 billion (FISHCOMM 1997).

Unlike Japan, which lost free access to waters around the world, the

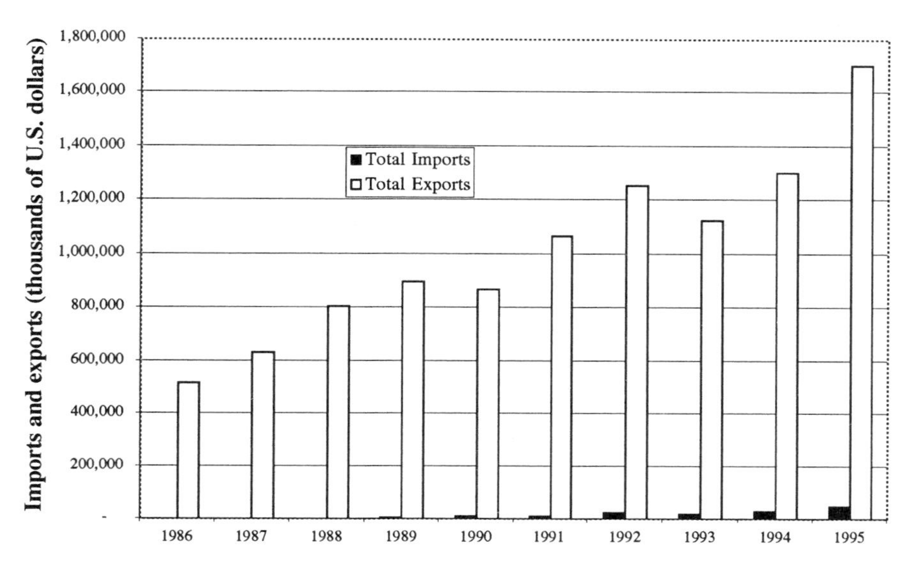

Figure 1.5. Chilean Imports and Exports of Fish, 1986–1995 (*source:* FISHCOMM 1997)

United States gained exclusive access to one of the world's largest fisheries—Alaska groundfish. After 1977, when the United States declared a 200-nautical-mile fishery zone around its shores, U.S. landings grew, from 2.9 million metric tons that year to 5.3 million metric tons in 1995. Still, the United States remained a net importer of fish. Imports of marine fish rose gradually from the mid-1980s to the mid-1990s, reaching $7.1 billion in 1995, compared with $3.4 billion in exports (see figure 1.6).

Fishing Effort

In assessing the sustainability of fisheries, fishery scientists and managers measure ability to catch fish in terms of fishing effort. In turn, fishing effort is often expressed as "number of days fished" or as size of vessel, such as gross tonnage. However, because fishing technology changes constantly, these measures offer an imperfect estimate of changes in fishing capacity. For instance, a twenty-foot dory powered by an outboard motor has greater fishing capacity than would the same dory powered by sail. Similarly, one fishing day by a trawler outfitted

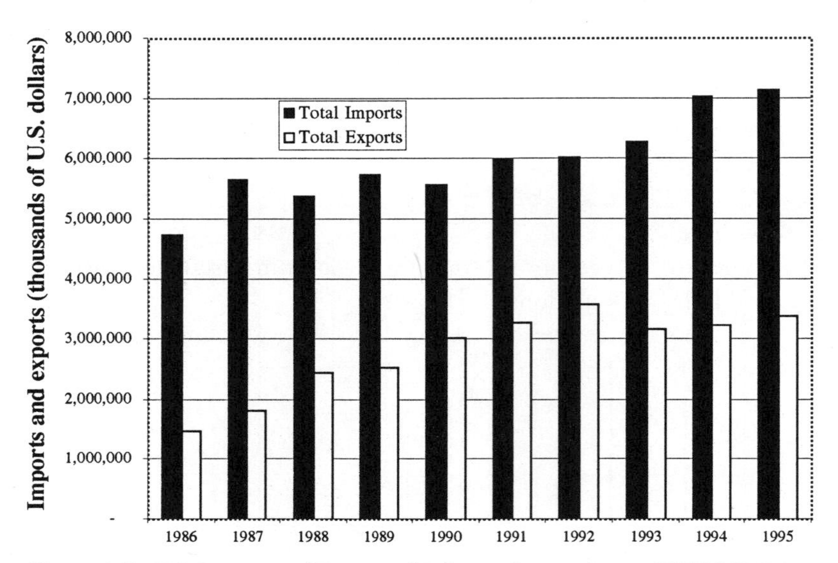

Figure 1.6. U.S. Imports and Exports of Fish, 1986–1995 (*source:* FISHCOMM 1997)

with a fish finder is greater than one fishing day by a trawler without one.

As it stands now, however, managers must make do with statistics on number and size of vessels when assessing trends in ability to catch fish. However, accurate statistics on fishing vessels are difficult to obtain. For instance, in 1992, Lloyd's Register of Shipping listed 24,400 fishing vessels of at least 100 gross registered tons (grt)[4] (generally considered the industrial fleet), whereas the *FAO Bulletin of Fishery Fleet Statistics* included 38,400 fishing vessels of this size (Garcia and Newton 1997). According to the Food and Agriculture Organization of the United Nations (FAO), fishers around the world use another 3.4 million vessels weighing less than 100 grt.

FAO statistics indicate that the total tonnage of the world's industrial fishing fleet increased by 4.6 percent per year from 1970 through 1989, moving from 13.6 million grt to 25.3 million grt (Garcia and Newton 1997). Fleets of developing countries grew especially rapidly, from 27 percent of the total number of fishing vessels to 58 percent. Measured in tonnage, fleets of developing countries increased their share of the world total from 13 percent to 29 percent.

On a global scale, expansion of fleet capacity has seemed to make economic sense. S. M. Garcia and C. Newton (1997) compared fleet size, landings, and value of landings per grt of fleet capacity for the years 1970, 1978, and 1989. Although overall fleet size grew by 87 percent, landings grew by only 46 percent. As a result, by the early 1980s, the world's fishing fleets as a whole were an estimated 30 percent larger than necessary to catch the maximum sustainable yield of the world's fish populations (McGuinn 1998).

This general trend toward overcapacity is reflected in individual fisheries, of course. For example, a 1991 study conducted for the Gulf of Mexico Fishery Management Council concluded that the area's offshore shrimp trawl fleet was twice its optimum size (Upton, Hoar, and Upton 1992).

Stagnating landings did not lead to decreased revenues, however. Lower supply and increasing demand raised prices so that revenue per grt actually increased from $2,100 to $2,300 (Garcia and Newton 1997). Higher prices contributed to continued, although slower, rates of increase in size and technological sophistication of fishing fleets in the 1990s (McGuinn 1998).

Status of Stocks

The difficulty and expense of studying populations of fish over large areas, together with the relatively low priority given to fisheries in scientific and government circles, has prevented assessment of many stocks of fish and shellfish in the United States and around the world. In some areas, such as the northwestern Pacific, more than half of the stocks cannot be assessed (Garcia and Newton 1997).

Nevertheless, in a recent review of 200 major fish stocks around the world, the Food and Agriculture Organization of the United Nations concluded that 35 percent of stocks showed declining yields, 25 percent had plateaued, and 40 percent were still developing (FAO 1997). As Garcia and Newton (1997) have noted, these assessments suggest there is little room for growth in landings. Indeed, they suggest that the percentage of stocks showing declining yields will increase if traditional fishing practices and management persist, for several reasons. First, so little is known about most stocks of fish that calculation of a maximum sustainable yield—a common objective of fisheries management—is highly uncertain. Second, maximum sustainable yield does not reflect a precautionary approach to managing many species of fish. Finally, as fishing fleets and fisheries grow, they develop an inertia that is difficult to deflect in response to declining yields. As a result, Garcia and Newton suggest, today's fully exploited fisheries are likely to be tomorrow's overexploited fisheries.

In general terms, fishers have maintained overall landings as well as landings of particular groups, such as tuna species, partly by moving from overfished areas to new areas (Grainger and Garcia 1996). For example, landings of demersal fishes such as cod and hake species peaked first in the Atlantic Ocean in the early 1970s, followed by the Pacific Ocean and then the Indian Ocean. Landings of demersal fishes in the northwestern Atlantic peaked in 1965 at 2.6 million metric tons; by 1995, landings had fallen to 331,862 metric tons (FISHSTAT 1997) (see figure 1.7). In the northwestern Pacific, landings from the world's largest demersal fishery peaked in 1986 at 7.4 million metric tons and then fell to 5.5 million metric tons in 1995 (see figure 1.8). With the decline in demersal fisheries in other oceans, demersal fisheries in the Indian Ocean have grown. From 1970 to 1994, landings

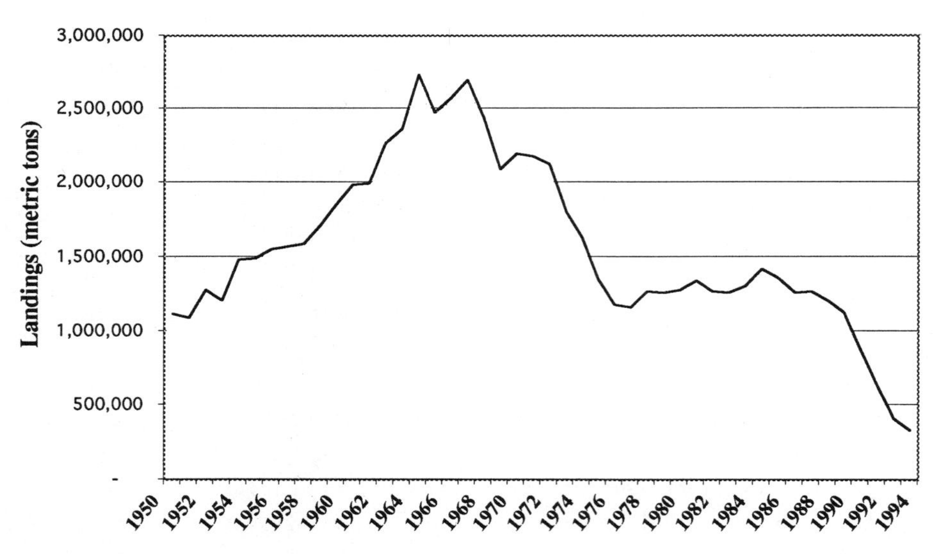

Figure 1.7. Landings of Demersal Fishes in the Northwestern Atlantic, 1950–1995 (*source:* FISHSTAT 1997)

in the western Indian Ocean increased from 372,350 metric tons to 943,673 metric tons (see figure 1.9).

Similar patterns have developed in fisheries for the principal market species of tuna—northern and southern bluefin, yellowfin, bigeye, skipjack, and albacore. As landings in the Atlantic Ocean have faltered, fishing effort and landings in the Pacific and Indian Oceans have grown. Overall, landings of tuna have been increasing by more than 7 percent per year (Garcia and Newton 1997).

In 1996, concern over economic and ecological dislocation associated with overfishing encouraged the United States Congress to require the National Marine Fisheries Service to report on the status of fisheries in U.S. waters (NMFS 1998). Specifically, the NMFS was to determine which species were overfished or approaching an overfished condition. Congress defined overfishing as a rate or level of fishing mortality that jeopardizes the capacity of a fishery to produce its maximum sustainable yield on a continuing basis.

Of the 844 species fished in U.S. waters in 1998, the NMFS was unable to determine the status of 64 percent—544 species—in the Caribbean region alone (NMFS 1998). Reef fishes and groupers in the

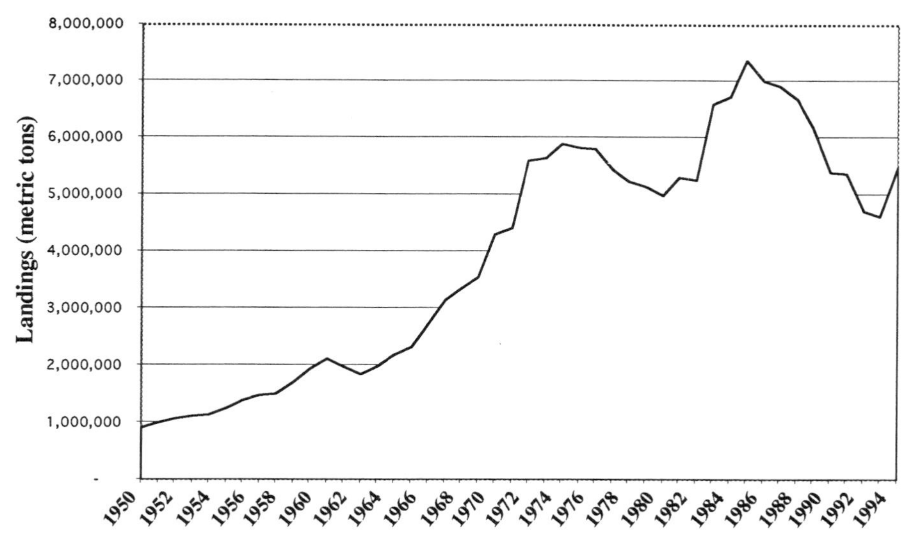

Figure 1.8. Landings of Demersal Fishes in the Northwestern Pacific, 1950–1995 (*source:* FISHSTAT 1997)

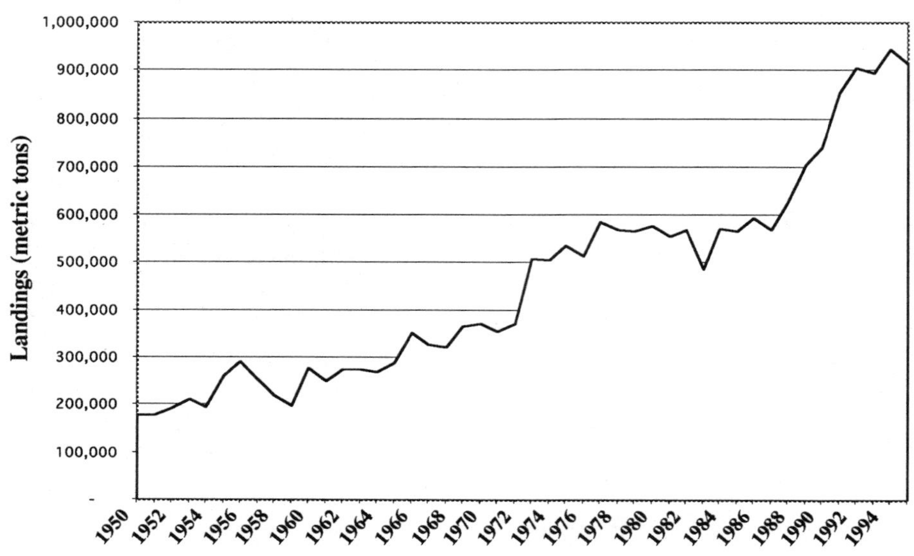

Figure 1.9. Landings of Demersal Fishes in the Western Indian Ocean, 1950–1995 (*source:* FISHSTAT 1997)

South Atlantic, the Gulf of Mexico, and the Caribbean Sea and rockfishes on the Pacific coast made up most of the species that were of unknown status and especially vulnerable to overfishing. Of the 300 species assessed, 90 were overfished and 10 were approaching an overfished condition (see table 1.2). The NMFS determined that the other 200 species were not overfished. Many of the overfished stocks are commercially the most valuable.

The NMFS also found that 11 other species not covered by fishery management plans were overfished. These included Atlantic halibut, tilefish, weakfish, spotted sea trout, spiny dogfish, Atlantic bluefin tuna, bigeye tuna, monkfish, Pacific sardine, and Pacific mackerel.

Around the world, governments and fishermen are confronting a very different task from that they faced in the 1950s. Then the world's fishing fleets were small and inefficient enough that they posed little threat of widespread overfishing. Marine fisheries offered a means to economic development and fulfilled nutritional needs. The buildup in fleet capacity, encouraged partly by government programs, overshot the capacity of many fish populations to produce surpluses. Now the

Table 1.2. Overfished Species Included in Fishery Management Plans Developed under the Magnuson-Stevens Fishery Conservation Management Act (FCMA)

Region or Category	*Species*
New England	Atlantic sea scallop, Atlantic salmon, American lobster, Gulf of Maine cod, American plaice, witch flounder, windowpane flounder, Gulf of Maine and southern New England winter flounder, southern Georges Bank silver hake, red hake
Middle Atlantic Ocean	Scup, summer flounder, black sea bass, bluefish
South Atlantic Ocean	Jewfish, vermilion snapper, red porgy, gag, red snapper, scamp, speckled hind, snowy grouper, warsaw grouper, golden tilefish, white grunt, black sea bass, red drum
Gulf of Mexico	Gulf king mackerel, red snapper, Nassau grouper, jewfish, red drum
Caribbean Sea	Nassau grouper, jewfish
Washington, Oregon, California	Chum salmon; summer and spring runs of chinook (king) salmon on the Columbia River; spring run of chinook salmon on the Snake River; spring, summer, and fall runs of chinook salmon on the Skagit River; summer and fall runs of chinook salmon on the Stil-laguamish and Snohomish Rivers; chinook salmon in Lake Washington and the Dungeness River; coho salmon in the Strait of Juan de Fuca
Western Pacific Ocean	Pelagic armorhead, squirrelfish snapper, longtail snapper
Highly migratory species in the Atlantic Ocean	Swordfish, blue marlin, white marlin, sandbar shark, blacktip shark, dusky shark, spinner shark, silky shark, bull shark, bignose shark, narrowtooth shark, Galápagos shark, night shark, Caribbean reef shark, tiger shark, lemon shark, sand tiger shark, bigeye sand tiger shark, nurse shark, scalloped hammerhead shark, smoother hammerhead shark, whale shark, basking shark, white shark

Source: NMFS 1997c.

task consists largely of reducing overbuilt fleets and rebuilding overexploited fish populations.

Notes

1. Much of this chapter is based on M. L. Weber, "A Global Assessment of Major Fisheries at Risk, Relevant Management Regimes, and Non-Governmental Organizations" (unpublished manuscript prepared for the Pew Charitable Trusts, 1998).
2. One nautical mile equals 1.15 statute (land) miles.
3. A gross registered ton (grt) is a measure of volume of storage space equal to 100 cubic feet.
4. A metric ton (or tonne) is a measure of weight equal to 1,000 kilograms or 2,204 pounds.

✦ Chapter 2 ✦

What Economics Has to Do with Fishing

In the introduction to this book, we proposed that the study of markets and economics could lead to a better understanding of how to manage fisheries. In the discussion of markets and economic incentives that follows, we present a view of fishing as an economic activity. Just as farmers bring wheat to market, fishers bring fish to market. Through the sale of wheat and fish in markets, farmers and fishers gain the financial wherewithal to sustain themselves and, they hope, to go out and produce wheat or catch fish again. The study of such markets, which have profound effects on farmers and fishers alike, is part of the field of microeconomics, which concerns itself with firms (in this case, the fishers selling their fish) and individuals (persons or other firms wishing to buy the fish).

Economics is the study of the allocation of scarce resources. By *resources* we mean things that people use.[1] Water, air, land, iron ore, wood, lumber, houses, and fish are all resources. In economics, a resource described as *limited* or *scarce* is scarce in relation to wants, or demand. In the list just given, iron ore, wood, lumber, and houses are of limited availability and therefore are scarce. Clean water is sometimes scarce and sometimes not, as is clean air. Likewise, fish are sometimes scarce and sometimes not. Farm-raised fish are scarce. Marine fish, however, have not historically been treated as a scarce resource.

Generally speaking, if the amount of a good that one person uses does not affect what is left for everyone else, it is regarded as *free* (as opposed to scarce). Under many conditions, there is unrestricted access to a resource and no one is charged a fee or is otherwise limited in taking it. Economists term such resources *open access resources* (as opposed to *free* goods). Such resources are not regarded as scarce, and therefore demand for them is potentially unlimited.

The function of markets and prices is to allocate scarce resources. *Allocation* is the determination of who gets what and how they get it. For example, in a tribal society, resources might be allocated by traditional rules or by tribal leaders. In communist societies, resources are allocated by central planners. In free market money economies, resources are allocated by means of *contractual exchange,* in which an owner of a resource freely agrees to trade it or sell it to someone else. This exchange takes place in a context known as the *market.* The price of products is determined in the market by buyers' willingness to pay and by producers' willingness to sell. When resources such as marine fish are open to access (that is, they are available for the taking) at the point of capture, one cannot contract for them, although, of course, once they have been captured they may subsequently be sold. They are therefore not subject to the market's allocating forces.

The Economics of the Farm

To understand how the market affects the allocation of scarce resources, consider a farmer who owns land and uses her own and hired labor, tractors, seed, fertilizer, and water to produce wheat. The farmer begins with what she possesses: her ownership of the land, her work capability, her knowledge of farming, her management skills, and money to hire labor and purchase the other resources she needs to produce wheat. The farmer will usually try to combine these resources—for example, labor, seed, and fertilizer—in the way that will bring her the greatest possible income. In economic language, the resources the farmer uses in production are called *inputs,* and the wheat crop is the *output,* or *product.*

For simplicity, assume that the farmer combines her inputs in order to achieve the highest possible income for her wheat. She is a *price taker* in both her input markets and her product market. That is, the price of both inputs and outputs is set in a market and she has no power to change these prices. She can accept them or not. Her first pound of seeds costs the same amount as the twentieth pound, and the first ton of wheat will fetch the same price as the twentieth ton. As one of many producers of wheat, she cannot affect the same price because one farmer's wheat substitutes for another farmer's. Nor can she carve a special niche by differentiating her wheat from others'. These assumptions are important for the graphs that follow.

The farmer will maximize her income by using her resources in the right combination. This means that in making her production decisions, she will consider her production costs in relation to the value of her output. For example, if one unit of fertilizer will boost production, and thereby income, by an amount that more than offsets the cost of buying and applying the fertilizer, the farmer will buy and apply it. She will, in fact, buy and apply more and more fertilizer until the cost of buying and applying another unit approximates the value of the additional output that can be anticipated from the use of the last unit of input. Economists call her decisions about buying fertilizer and other inputs *marginal cost production decisions.*

The cost of the additional unit of output is termed the *marginal cost.* For the farmer in this example, the way it changes in relation to the amount of fertilizer purchased is shown in graph (a) in box 2.1. Among economically efficient producers, marginal costs of production are less than or equal to the marginal value of the product. The value of each additional unit of input in terms of output is called the *marginal value of the product.* The trajectory of the marginal product curve, as shown in graph (a), is determined by the combination of inputs used to produce the product. Graphed over a range of input use, the marginal value of the product typically rises at an increasing rate over some part of the range and then rises at a decreasing rate as the effect of more inputs diminishes. When too much of the input is used, output decreases and so does the marginal value of the product.

At the level of input use at which the last input applied just barely

Box 2.1. Marginal Costs and Marginal Revenues

Marginal Product is the extra output obtained from an additional unit of input used in a production process. The production effect of the first unit of input applied is different from the effect of the tenth unit applied. Marginal Revenue Product (MRP), or Marginal Value Product, is the marginal (physical) product multiplied by the marginal revenue earned from the sale of the extra unit of output. It is the addition to revenues from producing more output from applying one or more unit of input.

Graph (a) shows total costs *(TC)* and total revenues *(TR)* for a farmer buying fertilizer. The total revenue curve is derived from the effect of the input (fertilizer) on the output (the farmer's crop). If the value of each additional unit of output is constant, the production effect can be measured in dollars. The first few units of input produce a slight increase in production. The rate at which fertilizer use adds to revenues earned changes at different levels of fertilizer use. Moving to the right on the horizontal axis, the effect increases more and more until after about 11 units (where the slopes of the TR and TC curves are equal) its effect on revenues is diminishing.

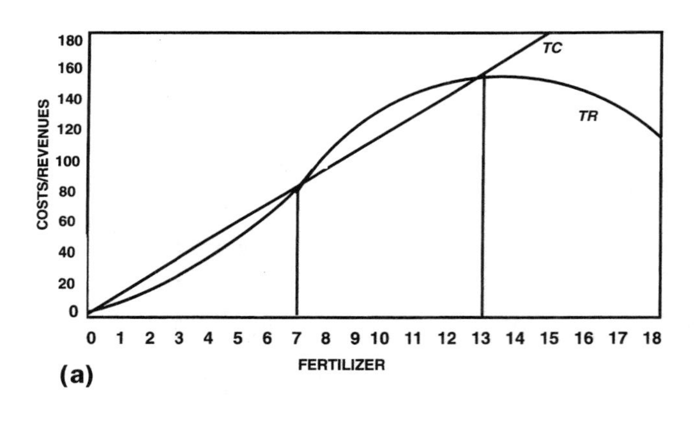

(a)

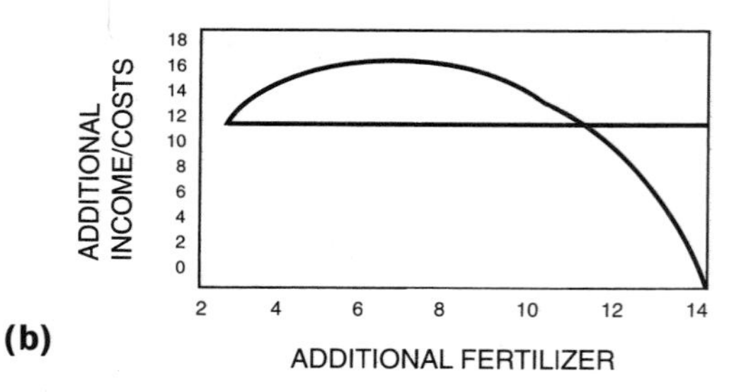

(b)

Graph (a) shows that from zero to 7 units of fertilizer, total costs are greater than total revenues. A farmer would not want to apply fertilizer at those rates. At about 13 units of fertilizer, the total cost curve again becomes greater than the total revenue curve, so the farmer would not want to use more than 13 units. But in the range of 7 to 13 units of fertilizer, would the farmer choose to stop applying more?

Graph (b) shows the cost of applying another unit of fertilizer, or the price of an additional unit. The marginal cost curve *(MC)* is flat because it does not show cumulative cost, only the cost of a single unit of input. The marginal revenue curve *(MR)* shows the additional revenue gained by using an additional unit of fertilizer. The effect on output is different at different levels of fertilizer use.

The marginal cost and marginal revenue curves show the point at which another unit of fertilizer will cost more than it is worth. At 11 units of fertilizer, the value earned by using another unit equals the cost of buying and applying that unit. This is the point at which profits are maximized. Beyond this level of fertilizer use, marginal cost is greater than marginal revenue and money is lost on each additional unit of fertilizer used. Back on the total revenue and total cost curves in graph (a), total revenue is still rising at 11 and 12 units of fertilizer. In that graph, it is not so obvious that revenue is rising less than is the additional cost of the eleventh and twelfth units of input. This is what makes the graph of what happens at the margin so useful.

pays for itself in terms of additional production, the profits of the farmer will be maximized because before that point producing more earns more revenues than costs, and after that point producing more costs more than it earns. Additional units of the given input would cost the farmer more than she would earn in revenue from the additional output. Competition among producers spurs them to combine their available resources so that the marginal cost of their last unit of production equals its marginal revenue in the market.

The idea that efficient producers equate marginal production costs with the marginal value of their output is a fundamental and useful economic concept. If an economist knows the shape of the production function (the mathematical equation that defines what we have just described with words) and input prices, he or she can predict the level of input use for any given market price. (Note that the production function alone does not determine the price, which depends equally on demand, discussed in the following section.) All other things being equal, higher output prices generate higher levels of input use. In the case of the fishery, using more inputs turns out to be an important part of the problem.

Paying the Costs of the Firms: The Demand Side

The amount of goods producers supply in any market depends on the prices of inputs and outputs. The amount of goods consumers demand depends on prices, alternatives, and consumption preferences. The eighteenth-century author of *Wealth of Nations,* Adam Smith, called the supply-and-demand phenomenon resulting from the self-interested behavior of producers and consumers the "invisible hand" of the market acting to allocate resources. When everything is working properly, the quantity supplied and the quantity demanded are negotiated in the market through prices, resulting in efficient allocation of resources.

Production efficiency, in economic usage, means that producers are using all their resources or inputs to get the greatest value. If the output quantity can vary, production efficiency is that level of production where profits are maximized (where the highest level of net revenues are being earned). This is where the greatest difference between total revenues and total costs exists, or where revenues equal marginal costs. Another measure of efficiency, used when the output quantity or goal is set beforehand, is *cost effectiveness.* This is where a given amount of output is produced at the lowest cost. The quantity consumers want varies with price. What happens is that under competitive conditions there will be a single price that will "clear" a market because at this price consumers' willingness to pay (demand) will equal buyers' willingness to sell (supply). This is where the demand and supply curves intersect. Under competitive conditions, demand and supply are matched by prices in markets.

Consider what would happen if an unexpectedly large harvest brought hundreds of extra tons of wheat to the market. If demand remained constant, the price of the wheat would have to fall in order for all available supplies to be sold. As an alternative, consider what

Box 2.2. Market Supply and Demand

Markets are where the free choices of many demanders and the self-interested responses of many suppliers meet at some mutually agreeable level of production. In this graph of price *(P)* and quantity *(Q)*, the demand curve describes a relationship between price and quantity. Higher quantities are wanted at lower prices; lower quantities at higher prices. Conversely, the supply curve shows that at higher prices, producers are willing to produce more of the good and at lower prices will produce less.

Here, the supply curve and the demand curve slope in opposite directions in terms of price and quantity, and they meet only once. The quantity that demanders would like to have is the same as the quantity that suppliers are willing to supply at one price. However, changes in consumer preferences or wealth can affect where the demand curve falls. Similarly, changes in technology, changes in input price, or both can affect where the supply curve falls.

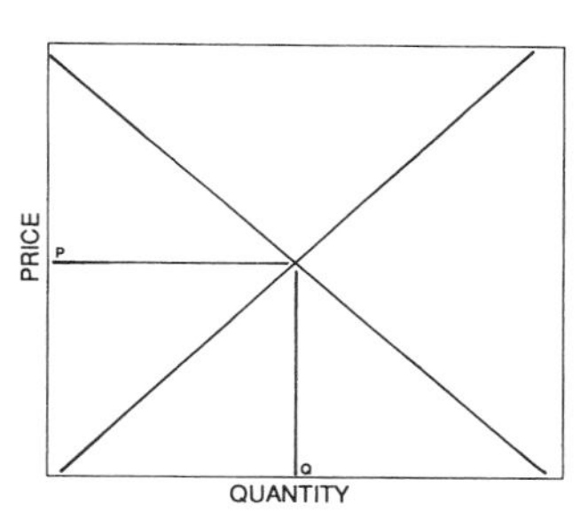

would happen if the Food and Drug Administration published a report saying that eating wheat cured cancer. Demand for wheat would increase sharply, prices would rise, and additional effort would be spent in bringing more wheat to market. In either case, price is the mechanism that would alert buyers to buy more or that would alert wheat producers to produce more.

Prices carry information from consumers to suppliers regarding how much they are willing to pay for a given amount of a good, and prices carry information from suppliers to consumers regarding how much can be produced at a given price. If consumers want more of a product than is being offered, they tend to bid up the market price to bring forth the additional supply. If suppliers want to move more of their products, they typically must reduce their prices.

Under this model of markets, based as it is on the allocative role of prices, market price plays a significant role in determining how much of society's available resources are used in making the product, and this in turn determines how much of the good is produced. This can be seen by noting that although suppliers are doing the spending by paying the costs of production, consumers give them the money to do so by buying the product. If consumers are willing to pay a higher price, more is spent on production. If consumers are willing to pay only a lower price, less of the good will be supplied or fewer resources will be used in its production. Given this, the area below the supply curve in box 2.2 reveals the amount of resources society is willing to spend on supplying the product.

Rents

When there is a positive difference between the cost of supplying an amount of a good and its market price, the difference is called *rent.* Rent is the value a productive factor has when it is used above the cost of using it. It is the residual after everything else is paid for, and it usually accrues to the owner of the productive resource. Land used in farming, for example, has a use value that, together with its scarcity value, helps determine how much is paid for it in the marketplace. On the farm, rents can be thought of as the profits that can be attributed to the land. Land rents will accrue over time (unless the land is degraded), so the value of the land is determined in part by the sum of net revenues that will come in future years. In a private property system, rents to land accrue to whoever owns the land. The owner can capture those rents by farming the land or, since rent is figured into its value, by selling the land.

Rents and ownership of the resource become quite important to our discussion of marine fish markets later in this chapter. It is important to understand the incentive effect of rents and ownership. When someone possesses a productive resource, that resource can be valued as a salable commodity in the market. If the resource is not being used, the owner is losing potential income. The owner thus has an incentive either to use a productive resource or to sell it. If no one owns the resource but it has value, it will be used until it generates no more rent for additional users.

In summary, in the simplest economic terms, producers use available resources efficiently to produce goods at a minimum cost, in amounts that consumers demand, at a price that will cover their costs of production and produce a profit. When wants and needs are satisfied by competing suppliers of goods and services in a free market, resources are efficiently allocated.

However, markets do not always work this way. Sometimes the assumptions that underlie this model do not square with actual conditions—especially in markets for marine fish.

The Problem with Fishing

The story of the wheat farmer, although helpful in illustrating some basic ideas from the theory of the firm, does not translate directly to fisheries. Why not? First, fishing in the ocean is quite different from farming or cutting timber on privately owned land. No one "owns" the ocean; it has traditionally been open to access by all comers.

Further, fishermen do not produce fish or harvest them; they catch them. Fish are known in economic terms as a *capture resource,* and the economic model for the way fish are brought to market is different from the production model just described.

As long as necessary environmental conditions prevail, fish produce fish. Fishermen catch fish. In the wheat production model described earlier, the wheat farmer owns the land on which the wheat is produced, and thus she is able to keep all the harvest to herself. If any person could come along and harvest her wheat before she brought it in, she would have little incentive to go through the effort of producing the crop. The farmer holds property rights to the crop by virtue of her ownership of the land.

Most basic to the nature of property ownership is the right to exclude all others. Since no one owns the ocean or the fish in it and anyone with the proper gear can catch fish from it, fishermen are, in some ways, like a row of competing combines lined up at the edge of the wheat field at harvest time. Without enforceable property rights, fishermen cannot be excluded from a fishery. They have an incentive to fish as rapidly as possible in order to prevent others from gaining access to the resource. What they do not catch, someone else will as long as there are profits to be made. This is the economic engine that drives open access fisheries.

Historically, the ocean has been *open access,* that is, without owners, without any set of rules governing its use and free to be used by all. Over time, parts of the ocean have become more like a *commons*—an area where owners agree to rules of use. For example, international law recognizes that countries have authority over activities, such as fishing, that take place within their territorial seas, generally twelve nautical miles from shore. Between the 1960s and 1980s, most countries extended their jurisdiction offshore to 200 nautical miles. The United Nations Law of the Sea recognizes a 200-mile Exclusive Economic Zone within which countries may regulate activity, however, from which they may not exclude the vessels of other nations. The high seas, areas outside the 200-mile zones of coastal states, are a *global commons,* where all nations have legal access. For a long time, natural resources within a commons were known as *common property resources,* but more recently, scholars have come to call them *common pool resources* (Ostrom 1977; Buck 1998). Once taken or used, common pool resources, including fish, are not available to others, and they have multiple users who are legally defined and difficult to exclude from the resource domain (Buck 1998; Ostrom 1996).

The idea of a commons comes from the feudal practice of setting aside grazing land for use by those who owned livestock but not pasture. The commons belonged to everyone in a particular village, and the value of the forage that grew on the commons was captured by whoever grazed animals there. Without rules for grazing agreed upon by all users, the pasture is open access. However—and this is Garret Hardin's reference to the "tragedy of the commons"—with open access to the commons for grazing, each villager has an incentive to graze larger numbers of livestock because doing so costs no more than grazing a few. Of course, the ever increasing numbers of livestock eventually overrun the capacity of the land to feed them all. Eventually the forage is an inefficiently allocated scarce resource, and everyone's welfare is diminished (Hardin 1968).

Marine fisheries are like commons because fish stocks have traditionally been available to fishers on a first-come, first-served basis. Just as forage on the commons generated value by fattening livestock, fish from marine fisheries generate value for fishers who capture them. And just as the herdsman faced an incentive to place as many livestock on the commons as was possible—even if doing so undermined the grazing potential for all the other livestock—the fisher faces an incentive to capture as many fish as possible even if doing so undermines the ability of the fish stock to maintain its abundance.

Such an incentive dominates these two cases because an individual herdsman's or fisherman's self-restraint without any control over what others do makes no difference to the outcome. When value can be captured from use of the open access resource, it is expected that if one person does not capture it, another person will. The value of the resource exists even after the land's carrying capacity or the fishery's sustainable yield has been exceeded; there is just less of the resource. Once the resource is overexploited, it is harder and more costly to catch, prices rise, and everyone loses, or, in economic terms, total welfare is diminished.

Why is there not such an outcome in the case of the farmer? Because she owns the farm and is able to exclude others from taking her wheat. In contrast, since no one owns the fish or the ocean, no entrepreneur will direct the application of inputs or effort in the most economically efficient way for the resource as a whole. Even as fish become more scarce and it becomes more and more costly to catch a dwindling number of them, as long as they remain to be caught, fishers will pursue them and the rent they represent. As the fish numbers dwindle, the price fishers can get for them increases, maintaining the incentive for additional effort and participants. Therefore, the rents that would, in the farm example, be taken by an owner (or forgone for future returns) instead remain to attract other participants. The addition of this extra effort not only reduces efficiency—wherein just enough catching power would be used to produce fish at the lowest cost—but ultimately results in the catching of more fish than the fish population can replace and sustain.

The farmer ensures that resources, such as new technology, are efficiently allocated in producing wheat from her farm. She will, predictably, take care of the farm in a way that will ensure future harvests. If she does not, she is wasting productive value, and people do not willingly waste value. In contrast, in fishing, as technology improves, the increased efficiency that should be increasing rents, or the economic surplus from fishing, actually dissipates because improvements in technology are what add too much fishing power or effort to the fishery, driving catches and stock size below the level a private owner theoretically would maintain.[2]

Marine fish stocks in an open access regime that does not exclude anyone tend to be overfished once the capability for overfishing exists. And the incentive is there to find the capability. As a population is fished down and the fish become fewer and harder to catch, fishers enhance their fishing capacity and employ better technology. Just as a farmer will apply fertilizer to eke out more production from depleted soil, a skipper will add more fishing power—faster engines, bigger boats, smaller mesh sizes in nets, larger trawl openings—as long as there are fish to catch. The inevitability of this outcome is described in greater detail in the following chapter.

In chapter 3, we show that fishers will pursue fish (or their resource value) even after they have passed the point at which the reproductive potential of the fish population can replace what is taken. Along the way, we describe the population dynamics of fish enough to support the idea that there is some amount of catch that, if taken year after year, is the maximum amount that can be taken without reducing the availability of fish in coming years (keeping in mind the caveat that maximum sustainable yield should be viewed as a ceiling rather than a target). We show that resource rents do not diminish to zero until this level of catch is surpassed. And because rents or profits are what attract people to fishing, more effort will be applied to fishing until these rents are driven to zero.

Complicating the Model

Until the 1960s, almost all the marine fisheries of the world were open access; that is, there was no proprietor or owner nor any regime to create responsibilities, duties, or property rights. Into this commons, fishers poured vessels, gear, and labor. Over time, with the onset of national fisheries jurisdiction and the evolution of fishery management, pure open access was modified. The present situation has been described as regulated open access (Wilen and Homans 1997) or controlled access (Sissenwine 1998). These systems combine controls or regulations with limits on access to the fishery.

If increasing rents, through improved technology or market demand, drew so much effort into a fishery that the catch threatened the fishery, managers and regulators stepped in with biologically based restrictions to protect the stock. These conservation controls could be gear restrictions, time or area closures, or effort limits, or they could be limits on total catch, size of fish, or number of fish per trip. Such restrictions can protect a stock for a while, but if access remains open, eventually they will not be enough to discourage new entrants from seeking the increasing rents in the fishery, with a concomitant increase in effort and catch.

As the demand for seafood increases—as it has continued to do with increasing human population, a rising standard of living in many countries, and a growing preference for seafood over other protein sources—increased investment in fishing is encouraged. It is this investment that caused catches to rise for twenty or thirty years before the 1970s. Since then, the absence of any rights regime or ability to exclude new entrants has caused catches to level off despite more and more additions of capital. If individual fishers were ensured rights in some portion of the catch and all fishers knew that the total catch was fully allocated, there would be no incentive for new fishers to enter the fishery or for existing participants to continue adding effort because the rents would already be entirely captured.

Restrictions on catch can protect a stock for a while, but management that ignores economic incentives and the importance of rights to fish tends to respond to disaster rather than avert it. If access remains open, catch limits and rules alone will not be enough to discourage new entrants from seeking rents in the fishery (Sissenwine and Rosenberg 1993).

In a predictable cycle, fishing power increases, stocks decline, managers institute tighter restrictions, fishers introduce new technology or fishing techniques, and operators apply their effort either in a new way in the old fishery or in some new (and vulnerable) fishery, causing those stocks to decline. In the cycle described in figure 2.1, increased competition motivates more stringent regulations, which in turn motivate more innovative and resourceful fishermen, renewed fishing power, and ultimately disaster.

Thus, if management of ocean fishery resources is to become more than disaster response, it will need to take into account the resource value of fish and the importance of assigning rights to those resources.

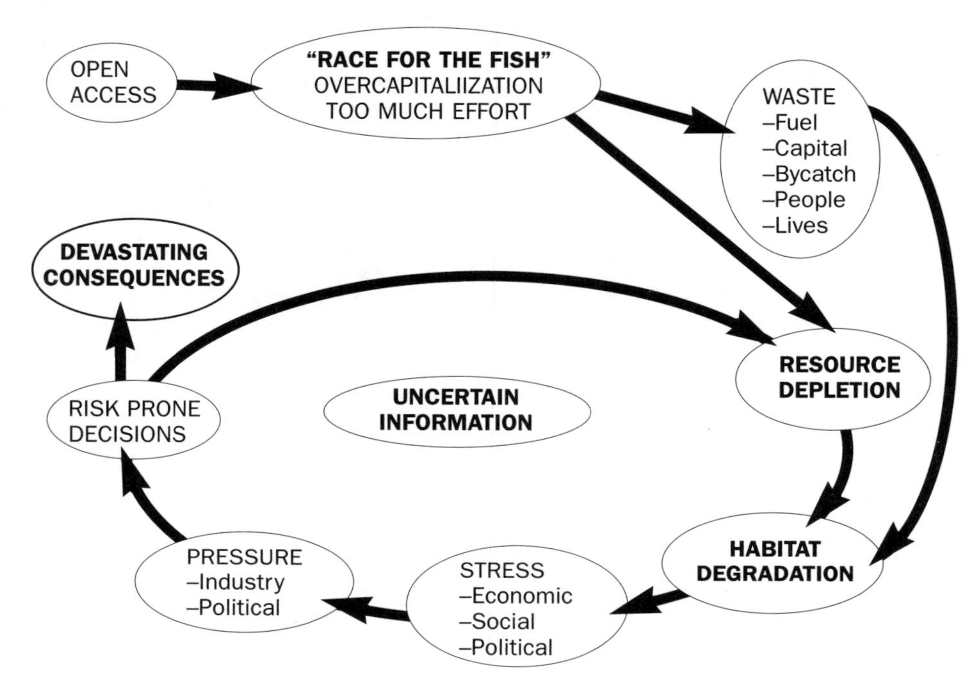

Figure 2.1. The Cause of the Problem (*source:* Michael Sissenwine and R. A. Rosenberg. 1993. "Marine Fisheries at a Critical Juncture." *Fisheries* 18(10): 6–11.)

Notes

1. This view of natural resources as commodities is an eighteenth-century Western way of thinking. Natural resources may be viewed otherwise in different cultures and in different disciplines, but we use the definition here for purposes of the economics discussion. See, for example, Hanna, Folke, and Mäler 1996.
2. The paradigm that attributes the cause of this behavior to open access and lack of ownership was first set out in 1929 by Jens Warming, a Dane (Warming 1983). His model was duplicated in 1954 by Scott Gordon. Since then, commentators have argued that it oversimplifies the situation by ignoring connections between economic rents, catches, and biology (Wilen and Homans 1997).

✦ Chapter 3 ✦

Biology, Economics, and Bioeconomics of Fish Populations

We made the assertion in the previous chapter that when marine fish stocks are treated as free goods in an open-access regime, they will be overfished if it is feasible to do so. In this chapter, we explain in greater detail why this outcome should be expected. It is not because fishers are greedy; nor is it because politicians and their appointees are inept. It results from the same incentives that push each and every one of us to get out of bed in the morning and go to work.

All of us want to make a living. The problem in marine fishing is that no one owns the fish, and their value in consumer markets is available for capture by whoever wants to do it. This chapter shows that it is the relatively unrestricted access to marine fish that makes fishers willing to fish stocks past the point of sustainable yield—past the point at which the stock can replace through growth and reproduction the animals lost to fishing mortality.

Yankee ingenuity took the New England groundfish fleet to the top of the landings chart in the boom years of the 1980s. Technology improved, and catches rose—to a point. As New England groundfish stocks declined over a decade or more, managers continued to clamp down on the fleet in order to protect the fish. They tried limiting the size of landed fish, limiting the size of mesh, and instituting short fishing seasons and small closed areas. However, with each regulatory stricture, skippers found a new way to change technology or improve their catching prowess, and stocks continued to decline. Finally, in 1994, days at sea were limited and large areas were closed to groundfishing. The federal government even financed a boat buyout in 1996 to reduce fishing effort.

The stocks are showing some recovery under a stringent rebuilding plan with severely reduced fishing effort, large closed areas, and conservative target catches. But observers would say this is just a pause in a cycle that will repeat again and again (Sissenwine 1998). Once the stocks show some signs of recovery, the pressure to ease up on limitations will increase, fishers will seek ways to avoid restrictions and improve their catching power, and the reproductive capacity of the fish will again be overcome by excessive fishing mortality.

The Biology of Fish Populations

Population dynamics is the story of how populations change over time. The study of any given population's dynamics entails looking at the reproductive strategy of the species in question, environmental factors such as food availability and habitat issues, and mortality factors such as predation and disease. In more complex cases, population dynamics entails the use of mathematical models and empirical data to explain and predict changes in populations over time. Such modeling provides a way to represent commonsense ideas in a framework that allows scientific analysis.

To introduce the population dynamics of marine fish stocks, we use the Schaefer logistic model of fish stock growth, explained in box 3.1. This model provides a working approximation of what happens among many important marine populations given different starting levels of stocks.

Although there is much more to the story, simply put, the curve produced by the model portrays graphically the notion that the num-

Box 3.1. The Schaefer Logistic Model of Fish Stock Growth

The Schaefer logistic growth model is described in the graph, in which the horizontal axis measures total population by weight, or biomass, and the vertical axis measures change in total population. For any initial stock size, measured on the horizontal axis, the annual production (or population growth) that can be expected from that stock can be read on the vertical axis. It will also be affected by changes in the carrying capacity of the ecosystem and a variety of other factors.

In other words, if you draw a straight line up from a given stock size, point *P*, to the growth curve, you can read the expected annual growth in stock size.

Note that the growth curve forms a bell shape. The same level of annual growth occurs for two different sizes of fish stock. At smaller stock sizes, growth is limited by the number of fish, whereas at larger stock sizes, growth is limited by environmental constraints and mortality.

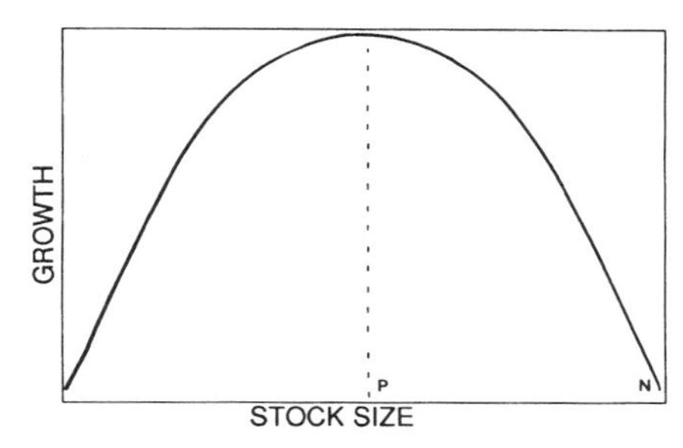

ber of fish a population starts with affects the number of fish that will be added to the population over a period of time. Because most fish are highly fecund (produce many offspring), a very small fish population in a high-capacity ecosystem can be expected to increase in number quickly. On the other hand, the increase in biomass, the total weight of all the fish in the population, would not be so great, since the population is starting from a small base and the recruits are all small, young fish.

In contrast, a population that is medium sized in terms of its ecosystem's carrying capacity can be expected to generate a greater increase in biomass per unit of time because there is still plenty of carrying capacity to support high growth. A population that is large in relation to its ecosystem's carrying capacity would grow less because of intraspecies competition for resources. The bell-shaped curve in box

3.1 illustrates that the greatest growth in biomass is generated by stocks that are medium sized in relation to carrying capacity.

The graph illustrates several important relationships. As one moves away from the origin (zero on both the vertical and horizontal axes) on the horizontal axis, the growth rate increases from one level of stock to the next until one gets to stock size *P.* Past stock size *P,* growth in biomass declines as stocks increase in size. At population size *N,* the population growth rate is zero. At some point (*N* on this graph) for some given species in some given environment, an equilibrium is established wherein new growth and mortality balance each other and the system is homeostatic. This stock level can be thought of as the natural equilibrium size. If some isolated event causes the population to fall—if, for example, an oil spill kills a large number of fish or changes in water temperature caused by an El Niño remove an important prey species—the population will respond by increasing each year at the rate implied by the growth curve until it reaches its natural equilibrium size. Fish populations larger than the equilibrium size have negative population growth over time because competition for prey, habitat, and other limiting factors in the ecosystem prevent them from growing indefinitely.

The biomass growth curve also demonstrates that there are two different stock levels at which the same rate of biomass growth occurs. At one of these points, biomass growth is constrained by biological productivity factors under which relatively large numbers of new fish are recruited and the existing stock grows well, but growth can be only so great because of the smaller starting population. At the other point, population growth is constrained by competition among fish, greater mortality, and so forth. At population size *P* in the graph in box 3.1, the growth rate is maximized. It is as great as it can be, given the ecosystem's carrying capacity and the population's life history characteristics. It is important to note that these relationships are greatly simplified for this explanation. The study of the population dynamics of fish stocks incorporates myriad variables, including the number of fish of a given age in the population, how fast the species reaches maturity, whether environmental conditions affect the availability of prey species, whether favorable spawning habitats or conditions are available, and so forth.

In the farming example from the previous chapter, more fertilizer, up to a point, results in more production. In a fish population, larger fish stocks, up to a point, produce more biomass. Past that point, however, larger fish stocks produce less biomass. The farmer decides how much she wants to produce on the basis of input prices, output prices, and the production effect of the inputs. In contrast, the number of fish available to be caught in a fishery is controlled primarily by biology. Fishermen affect this response by taking fish out of the population, and they affect it to different degrees depending on whether they target large, mature fish, spawning females, or fish across a number of age classes. The mortality caused by fishing is added to all the other variables that can change the status and abundance of the population. The resulting changes in level and composition of fish stocks affect the population's ability to grow and reproduce. These effects are largely uncontrollable in an open access arrangement, wherein anyone who wants to capture the value of the fish is free to try.

Fishing and Population Biology

When human fishing effort is incorporated into the Schaefer logistic growth model, a new source of mortality affecting population size is introduced. When fishing effort reduces a fish population, the annual replacement of biomass changes, depending on the extent of the reduction, on the size of the starting stocks, and on all other things remaining equal. If all other things are expected to remain equal (and again, this is a simplification), the Schaefer logistic growth model can be used to generate a catch curve that shows the annual maximum sustainable catch for different levels of fishing effort.

The catch curve described in box 3.2 shows the maximum sustainable catch per unit of fishing effort. It is important to understand that each level of effort on the horizontal axis is taken to imply a certain (different) equilibrium population size. That is, for any given catch capability (fishing effort), there will over time be a corresponding population equilibrium. Greater effort will reduce the natural equilibrium of the exploited stock; less effort will leave it larger. Because of this assumed relationship between effort and long-term stock size, the shape of the catch curve is determined by the natural replacement that can be expected from a specific stock size, which itself is a result of the long-term effect of that level of effort. For any level of fishing effort E,

Box 3.2. Maximum Sustainable Catch Curve

The graph shows the relationship between different levels of effort and catch and the maximum sustainable catch that can be expected at any of the given effort levels on the horizontal axis. It is derived from the Schaefer logistic model of fish stock growth, described in box 3.1.

The maximum quantity of fish that can be caught in a year without reducing the stock size is given by the catch curve. This curve is determined by the biology of the fish stock and the environment. At an effort level of zero, no fish are caught. As effort increases, population size can be expected to decrease.

At lower effort levels (those to the left of *E*), the maximum replacement described by the catch curve is limited by the fact that the population is still very large in relation to the carrying capacity of its environment. Thus, very light effort will not take many fish, but the fish stock will not have a fast growth rate, either. At higher levels of effort (up to effort level *M*), larger annual catches are sustainable because limits on the environment's carrying capacity become less of an issue. Past effort level *M,* however, additional effort will reduce catch levels. At those levels of effort, so many fish are taken that growth is limited by smaller stock sizes. Thus, *M* is the level at which the fishery experiences maximum sustainable growth.

All points on the curve are sustainable, but if catch exceeds any point on the maximum sustainable catch curve, the fish stock becomes depleted. Conversely, if catch is below the maximum sustainable catch curve, the population moves back toward a larger size.

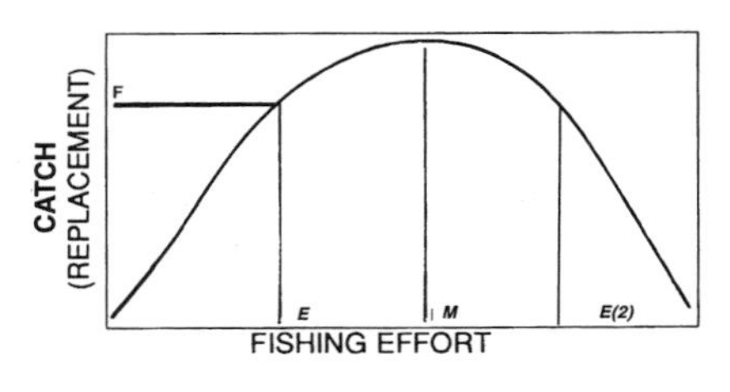

fishing mortality can occur at *F* amount or less of fish in perpetuity, all other things remaining equal.

Just as the Schaefer logistic growth model shows that the same amount of production can be expected for two different levels of stock, the graph of effort versus catch shows that the same level of sustained catch can be had for two different levels of fishing effort, one on each side of the line *M,* at which the catch per unit of effort is the highest. One must be careful with this, however, because stock size (on the

Schaefer graph) does not substitute for fishing effort (on the catch curve graph). On the catch curve, the stock scale is reversed. Smaller populations will require more effort to obtain a certain annual catch. That is, past where the curve intersects *M,* the population gets smaller. Larger populations, those to the left of *M,* allow a given amount of fish to be caught with less effort. Thus, lower levels of fishing effort imply larger stocks of fish, and higher levels of effort imply smaller stocks of fish.

If one can get the same catch from two different levels of effort, it makes sense to fish at the level that requires the least amount of effort and to regard the other side of the curve as irrelevant. Unfortunately, as will be shown later, an open access fishery tends to lead to the less desirable side of the graph.

To recap, both very small and very large fish populations generate small rates of annual biomass growth. Medium-sized populations generate larger amounts of annual growth. One population size will produce the greatest possible increase in a year, called the maximum growth rate. However, any given stock size will have a maximum sustainable catch rate—an amount of fish that can be taken out of the stock without reducing the amount that will be available in the following year. Thus, catch depends on effort and fish stocks, and except for one certain level of effort (the maximum sustainable catch rate), any given level of sustainable catch can be had at two different levels of effort.

So why do fishermen spend more effort than needed to catch fish?

Economics of the Open Access Fishery

In the previous two sections, we described how it is possible to project growth in fish biomass among marine fish stocks given different starting levels for the stocks. We also showed that this growth model can be combined with expectations about equilibrium stock sizes under different levels of effort to determine the maximum sustainable catch for any level of continuous fishing effort.

If the reasonable assumption is made that the price of fish will remain constant over all the different catch levels graphed in box 3.2, it is clear that this price can simply be multiplied by each maximum catch to determine the total value of different levels of catch. This produces a total revenue *(TR)* curve, as shown in box 3.3.

Moreover, if it is assumed that for the fishing industry as a whole, an extra level of effort requires a constant extra cost—for example, if one more boat-day costs an additional $1,000—total industry costs can be represented on a graph of money versus effort. When total revenues and total costs are represented on the same graph, the groundwork is laid for determining the economically optimal level of fish catch. This is, of course, an oversimplification. Boats cannot be substituted one for another—the wide array of vessels, the diversity of expertise of their skippers, the facility of crew members in deploying and hauling back gear, and the amount of time spent fishing all contribute to the total effort.

The graph in box 3.3 (in which the total cost, or *TC*, curve intersects the *TR* curve to the right of *M*) shows that depending on the relative placement of the curves, total revenues do not drop below total costs until well past the level of effort that provides the greatest biological yield. This plot illustrates why fishermen in an open access fishery will tend to spend too much effort chasing too few fish. Taking any point on the *TC* curve to the left of where it intersects the *TR* curve, it is clear that the level of effort corresponding to that point could generate more revenue (draw a vertical line from the *TC* curve to the horizontal axis and to the *TR* curve) than cost. The piece of a line drawn from the *TC* curve to the *TR* curve can be thought of as industry profits.

Consider the farmer again. If another unit of fertilizer could generate enough extra revenue to pay the cost of buying and applying the fertilizer and still leave an additional amount for the farmer, she would choose to apply the fertilizer. But we showed in box 2.1 that total revenues could still be growing even after marginal revenues began to decline. In the example of wheat production, we said that the farmer gets to choose how much fertilizer to use and that she stops at the profit-maximizing point, where marginal costs equal marginal revenues.

A fisherman will also strive to maximize private profits. He will combine the elements of his effort—vessel, gear, labor, fuel, time on the fishing grounds—where there is the greatest excess of revenues over cost (to the left of the maximum catch point in box 3.3). He will continue to apply these inputs as long as his return from the catch exceeds

Box 3.3. Total Costs and Total Revenues in the Industry

The graph shows the catch curve factored by the value (price) of the fish. The catch curve, showing the relationship between effort and maximum sustainable catch, thus becomes the total revenue *(TR)* curve. The line from any point on the effort axis to the *TR* curve gives the maximum sustainable revenues that could be extracted from the fish population at a given level of effort. At effort level *E*, the maximum sustainable industry revenues would amount to *P* dollars.

The graph also shows the total costs *(TC)* for the industry. As with the farmer considering the effect of a single input in her production process, an additional unit of effort adds a constant cost to the industry. Thus, the *TC* curve is a straight line.

Maximum sustainable total revenues are earned at point *M*, where total costs are the farthest below the total revenue curve. All points on the revenue curve are sustainable, but at only one point does the fisherman earn maximum sustainable net revenues (profits), at point *E*. Point *I* shows the sustained revenue under open access, where costs are the same or more than revenues.

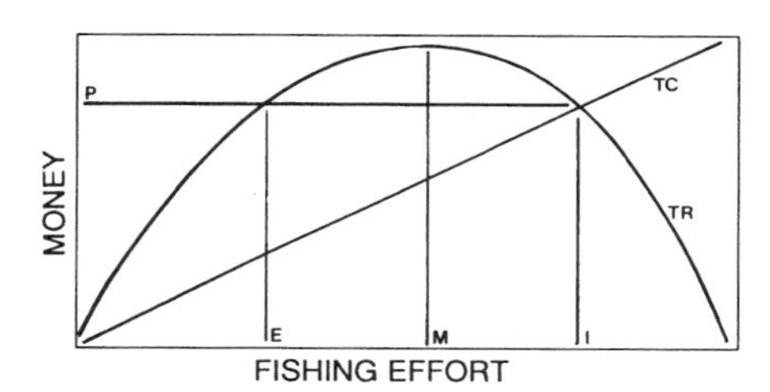

the cost of the inputs. In an open access fishery, no one places any limit on how much fishing effort is applied, and this is the root of the problem. If Skipper Uno adds several panels of net, Skippers Dos, Tres, and Cuatro will have an incentive to do so as well because if they don't, Skipper Uno will get the fish before they do. If Captain Cinco adds the net panels and installs a larger engine so his boat can get to the fishing grounds faster, others will also be moved to do so. As long as fishing generates profits, additional effort will be applied toward catching fish. The end result is that fishing effort will increase until there is no more profit to be made, that is, until the cost of the effort outstrips the revenues. And at that point—if what we have said about growth, production, and total revenues is true—too much effort is being expended in chasing too few fish.

A Closer Look at Industry Costs, Revenues, and Fishing Effort

In this section, we look closely at an important presumption that was slipped into the final piece of the model of an open access fishery. By changing that presumption, we show how technological improvements change the economics of a fishery.

One factor in determining total costs of fishing is the efficiency of fishing technology. In box 3.3, the total cost *(TC)* curve presumes a certain level of technology. However, what if the technological capability were not so great in relation to the task of fishing this hypothetical stock? In other words, what if the total costs for catching a certain amount of fish were to increase and the *TC* curve rose sharply and intersected the *TR* curve to the left of the maximum sustainable yield, as in box 3.4?

A fishery in which total costs equal total revenues is not economically efficient, as we showed earlier with the example of the farmer.

Box 3.4. Another Look at Industry Total Cost and Total Revenue

Historically, fishing fleets have used technology incapable of landing the maximum sustainable level of catch. The graph shows the outcome such technology achieves. Since the costs of using this technology are relatively high, the total cost curve *(TC)* rises steeply, intersecting the total revenue curve *(TR)* to the left of its peak. Under open access, even if the fishery does not deplete the stock, it is not efficient. The incentive to fish away rents in the fishery will exist as long as profits are possible. In response, individual fishers will apply fishing effort until total revenues in the fishery as a whole equal total costs. On the graph, this economic equilibrium occurs at the point where the total cost curve *(TC)* intersects the total revenue curve *(TR)*. Here, where fishers are applying fishing effort at level *G*, both total costs and total revenues are equal to *P*. The economically optimum level of effort, where net revenues are maximized, is at point O.

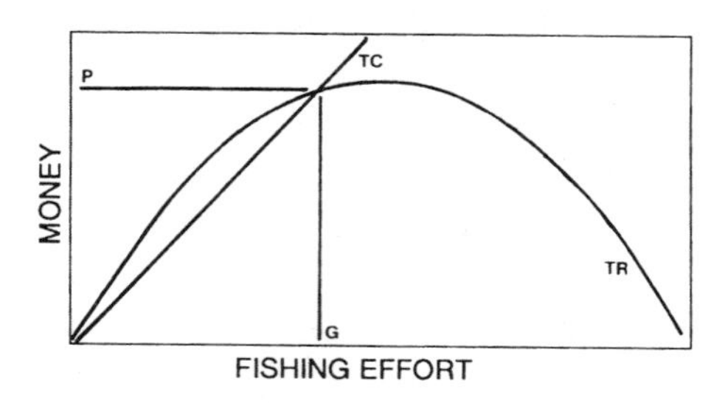

This is because marginal costs will exceed marginal revenues when total costs equal total revenues. Put another way, when total costs equal total revenues, the additional cost of catching another fish exceeds the additional revenue that fish will fetch.

Box 3.4 shows a fishery in which the fleet uses technology that is incapable of landing the maximum sustainable level of catch. Since the costs of using this technology are relatively high, the total cost curve *(TC)* rises steeply, intersecting the total revenue curve *(TR)* to the left of its peak at maximum sustainable yield.

As the graph indicates, this fishery is incapable of catches that will result in depletion of the stock but it is also not efficient. Assuming that access to the fishery is open, the incentive to fish away rents in the fishery will exist as long as profits are possible. Responding to this incentive, individual fishers will apply fishing effort until total revenues in the fishery as a whole equal total costs.

Many fisheries around the world that use conventional technology have operated for centuries at this level, where total costs equal total revenues. Although these fisheries may not produce maximum economic yield, they do not expend more effort than necessary to catch the available fish. However, the use of conventional technology does not preclude overfishing, as river and nearshore fisheries in many parts of the world attest.

What if more efficient fishing technology is brought into play? Three things can happen when improved technology is introduced into a fishery:

- Fishers may be able to expend the same level of fishing effort at lower cost. For instance, the use of synthetic line rather than cotton or hemp may allow fishers to make larger nets that can cover larger areas and require less fuel to drag through the water.
- Fishers may be able to expend more fishing effort at the same cost.
- Fishers may be able to expend more fishing effort at lower cost.

Whichever of these alternatives a fisher exercises, the cost-effectiveness of fishing will increase. In terms of the graph, the fishery's total cost curve will pivot downward, as shown in box 3.5.

Box 3.5. Industry Total Cost and Total Revenue with Changed Technology

The graph shows what happens with a change in technology from that shown in box 3.4. The new technology (represented here in bold type) produces a larger catch at a lower level of effort and lower cost per unit of effort. In terms of the graph, the new technology shifts the total cost curve ***(TC)*** downward. The net effect is that the total cost curve now intersects the total revenue curve to the right of the peak in sustainable revenues. At this point, not only is the fishery biologically inefficient, since catch levels are below the maximum level, it is also economically inefficient, since costs consume revenues entirely. Furthermore, the fleet is expending greater effort for reduced revenues. In the graph, revenues have fallen from *P* to ***P*** and effort has increased from *G* to ***G.***

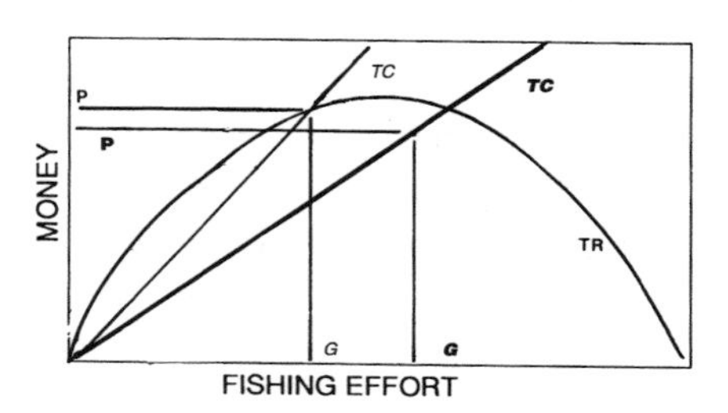

The introduction of improved technology affects not only costs but also revenues. Since more fish can be caught for less effort, a unit of effort with the new technology will produce more revenue than a unit of effort with the old technology. Total revenues will increase more rapidly toward the same level but will be achieved at a different level of effort. In terms of the graph, the total revenue curve will become steeper, just as the total cost curve became shallower.

The other side of this coin is that total revenues will decrease more rapidly once effort and catches have exceeded the maximum sustainable yield. At some point, the increase in catch made possible by technological improvement will exceed the population's rate of growth. At the new equilibrium, more effort will actually catch fewer fish because of the resultant decline in the fish population. There is considerable empirical evidence that this is just what has happened in many important ocean fisheries. (In terms of the graph, the fishery will tend to increase effort with the new technology until the point of intersection of the total cost curve with the total revenue curve has moved to the right of the peak—the maximum sustainable level of biological yield and of revenues.)

An improvement in technology can increase the profitability of fishing over a limited range of fishing effort. Within this range of effort, the difference between revenues and costs will increase more rapidly than under the old technology, providing increased profits, and more fish will be supplied to consumers. If fishing effort could be confined within this range, more fish could be brought to market at a lower cost.

This range of effort, which produces maximum profits, will be below the level of effort that produces maximum total revenues. In terms of the graph, this range of effort is to the left of the peak in the total revenue curve. However, as we have shown previously, fishing effort in an open access fishery will not remain within this range but will increase until all profits have been dissipated.

The Sole Owner Fishery

If the fishery belonged to a single owner who could exclude other fishermen, the outcome would be quite different. Under sole ownership, the fisherman would fish at the level of effort that maximizes profit. But unlike the fisherman in the open access fishery, where he—and all other competing fishermen—will continue applying effort as long as the return on a unit of effort exceeds its cost, the skipper in a sole owner fishery would have other options to maximize his return.

Such a skipper would be concerned about the number of fish he could catch in the future as well as those he might catch today. If he could exclude others, he would not have to worry about leaving fish for others to catch. A similar choice by a skipper in an open access fishery would not conserve fish for the future because as long as there were profits to be had from fishing that stock, someone else would catch right now whatever the forward-thinking fisherman might leave for the future. The skipper in the sole owner fishery might choose to fish for only the largest animals or to take the time to sort for the best-quality catch. Without the need to catch the fish before someone else does, the sole owner would not have an incentive to use more fuel, gear, and labor than necessary. Because the sole owner could exclude other skippers, he could choose the profit-maximizing level of catch, which is beneath the maximum sustainable yield.

An Alternative View

It is useful at this point to take a small side trip out of the economics paradigm and examine the view from other disciplines that contribute to an understanding of fishing behavior and fishery management today. Thus far, we have viewed fishing behavior from two aspects only: biological, or protection of the fish in the ecosystem (conservation), and economic, or maximization of returns on the use of the resource (rationalization). But people are not simply rational economic beings; nor are they entirely in harmony with nature, recognizing their dependence on the environment's productivity over the long term. People are social creatures as well.

A major criticism of Garrett Hardin's "tragedy of the commons" parable is that it assumes people are individualistic and greedy and their use of natural resources will lead inevitably to depletion (Wilen and Homans 1997). The model ignores social behavior and the possibility that people might act in concert to establish rules and standards for use of common pool resources (Hanna and Jentoft 1996). In the absence of self-control, according to Hardin's thesis, the self-interested behavior of resource users can be controlled only by government. At the other end of the spectrum is the view that the self-interested behavior of resource users demands privatization, wherein their actions will be regulated by market forces (McCay 1996).

Social scientists posit a third possibility: that there could be a wide range of property rights regimes, from those in which there are no rights, as in an unregulated, open access system, to the case of a sole owner of rights, or a private property system, in which the individual has the right to exclude others. Ownership could be collective, shared by all citizens, as in state-owned or public property systems, or could entail combinations or degrees of all ownership regimes (Hanna, Folke, and Mäler 1996). Further, the governance of an activity, such as fishing, within such ownership regimes could range from no management to market regulation to communal governance to state governance (McCay 1996). The mix of governance and property regimes could vary, depending not only on biological and economic goals but also on social goals such as job creation, local control, or preservation of community culture and heritage.

That said, the sole owner scenario is useful in illustrating how catch outcomes would change if one of the fundamental assumptions of the model were different. This particular change—moving from many self-interested fishermen to one—allows us to state simply the economically efficient best case for a fishery. That is the profit-maximizing outcome. It is worth considering whether government, to the extent that the public entrusts management of the resource to the government and its agents, acts as a sole owner. Many governments have acted in the conservation mode using regulation to constrain catches. A few have adopted the economic rationalization mode, shifting to property regimes that confer access rights or quota rights to individual fishermen, who then have incentives of a sole owner. We make the assumption that whoever the sole owner is, the objective will be maximization of the returns earned by utilizing the resource, as illustrated by the two graphs in box 3.6.

Profit is the difference between costs and revenues. Therefore, profit maximization, as shown in graph (a) of box 3.6, is the level of effort that produces the largest difference between the *TC* curve and the *TR* curve when total revenues are greater than total costs. That level of effort is achieved at point *D,* which is to the left of maximum sustainable yield. The point of maximum profit at the left of maximum sustainable yield has some intuitive appeal because at larger stock levels (which is what lower levels of effort imply), fish are easier to find and the relative cost of catching them is reduced. The economically efficient long-term catch in a sole owner model will, under most conditions, appear on the graph to the left of maximum sustainable yield, the point at which there are lots of fish and lots of profits to be had from catching them.

Although theoretically the sole owner scenario can be posited as the optimal, or most economically efficient, outcome that can be achieved in the fish market, it does not exist in the real world any more than does a completely unregulated open access regime. Most commentators would agree that today's fisheries operate under a combination of regulated open access, or controlled access in combination with regulations.

Box 3.6. Optimal Outcome for a Sole Owner

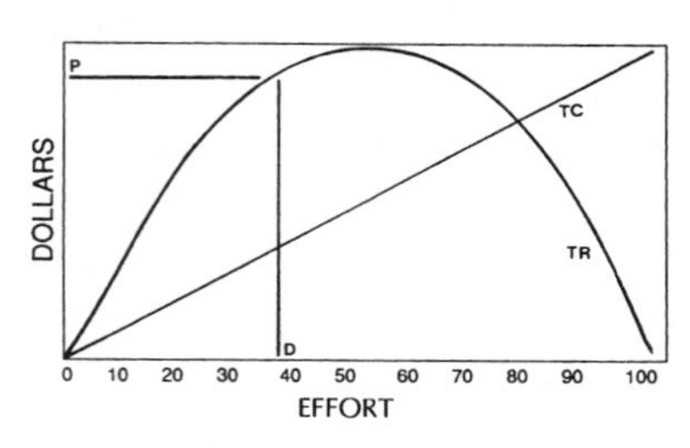

Graph (a) shows much the same story for the sole owner fishery as did the graph of total revenue and total cost for the individual farmer, except that here the input is fishing effort. The profit-maximizing level of effort can be seen as the maximum difference between total costs and total revenues, the effort level at point *D,* as long as revenues are greater than costs. If certain conditions are met, the maximum difference between costs and revenues is always at the point where the slope of the *TR* curve is the same as the slope of the *TC* curve.

The marginal revenue *(MR)* and marginal cost *(MC)* curves in graph (b) derive their shape from the *TR* and *TC* curves in graph (a). The *MR/MC* graph illustrates that the greatest marginal revenue occurs at the same point as maximum net revenues in graph (a). This is the optimal amount of effort. The marginal cost curve shows the cost of each additional unit of fishing effort. At a point beyond optimal effort, just before 40 units (or point *D* on graph [a]) the revenue from each additional unit does not cover the cost of applying that effort.

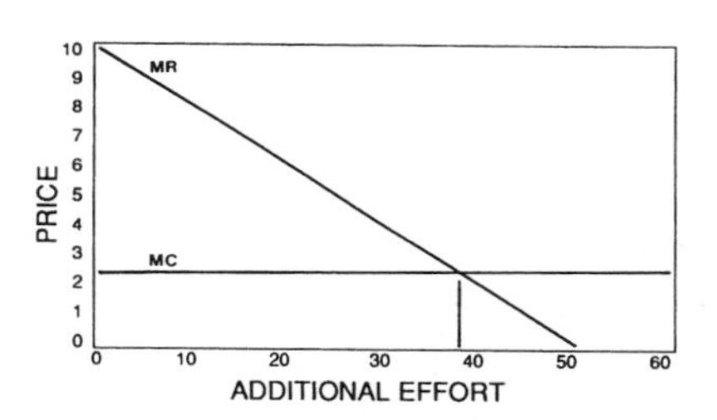

An understanding of the incentives in an open access system, however, is beneficial in examining some real-world policies that have been proposed or applied to reduce the overfishing brought about by modern technology and economic incentives in open access fisheries. The sole owner outcome serves as a standard against which to measure these other policies, to demonstrate what sort of information would be necessary to institute each policy and some of the enforcement problems each might entail.

Overfishing has seriously depleted populations of species such as bluefin tuna, sturgeon, swordfish, and some reef fishes. This is partly because improved technology and high prices have caused the total revenue curve to become very steep indeed. For example, sashimi-grade Atlantic bluefin tuna weighing 700–800 pounds frequently fetch $10,000 or more at the dock. Between these prices and the sophisticated technology used in finding Atlantic bluefin compared with that in use as recently as the mid-1970s, it is no surprise that fishing effort remains high in this fishery and that fishers resist reductions in quotas, even though the stock has declined by 90 percent since that time.

The next chapter describes how government policies—in the form of subsidies—have actually encouraged the overcapacity of the fishing industry. Many such policies have canceled out government efforts to control overfishing because they exaggerate the sets of incentives described here.

✦ Chapter 4 ✦

The Effect of Subsidies

From an economic perspective, the goal of fishery management efforts, if not the means, is clear.[1] To maintain an efficient allocation of resources and sustain profits, incentives to fish must be brought into line with the biological productivity and ecological viability of fish populations. In the real world, however, a great many other objectives conflict with the economically rational one.

Over the past several decades especially, governments around the world have played two often conflicting roles in marine fisheries. In the first role, governments typically have sought to secure food supplies, to increase employment, and to promote economic development by increasing the capacity of their fishing fleets and processing facilities. In the second role, governments have tried to protect society's interests by restraining fishing fleets from depleting public fishery resources. It is the first of these two roles with which this chapter is concerned.

Until the late 1980s, the former role dominated as governments promoted social objectives through economic assistance. In 1639, for instance, the General Court of Massachusetts rewarded cod fishermen with exemptions from military duty and from taxes on their gear and vessels (Weber and Gradwohl 1995). This policy was aimed at encouraging an industry that generated much of the colonies' wealth and resources for the American Revolution. Economic assistance to this fishery continues to this day, even though it no longer plays nearly so significant a role in the nation's economy.

A subsidy can be defined as a government policy that alters market risks, rewards, and costs in ways that favor certain activities or groups (Roodman 1996). Several things should be noted about this definition. First, subsidies are not simply tax breaks or grants; they may include other kinds of government policies, such as those involving research and development or marketing.

Second, the key feature of subsidy policies is that they distort the way markets operate, allowing some groups to benefit over others. Inevitably, disadvantaged groups succeed in securing the benefits of other subsidies in order to maintain their parity in the marketplace. Results include the pervasiveness of subsidies in modern economies and the difficulty of knowing where to start in reducing them.

Subsidies have promoted overexploitation of marine wildlife, just as they have promoted overuse of forests and other natural resources by distorting the conditions under which investors, fishermen, processors, and consumers make decisions. First, subsidies can attract more private investment in exploitation of natural resources than would otherwise occur (Porter 1997). Some agricultural and fishing subsidies, for instance, have led to more marginal land in cultivation and more fishing boats. Second, subsidies may encourage the use of technology that increases the capacity to exploit natural resources. Third, by reducing the prices of natural resources such as fish populations, subsidies discourage their efficient use.

Subsidies that reduce the marginal cost of exploitation, as through reduced fuel costs, encourage higher levels of exploitation than would occur if these costs were not artificially reduced. By reducing or eliminating fees for the use of a public resource, such as trees or fish, subsidies also encourage higher levels of consumption. In terms of the graphs of total cost and total revenue in chapter 3, the effect is to flatten the trajectory of the cost curve, thus maintaining the incentive to continue to fish even further past the point at which revenues begin to decline and further past the point at which fish stocks can replace the fish caught.

In addition, the decision to charge little or nothing for the use of public resources for private gain is a decision to forgo revenues that government could devote to necessary research, monitoring, and surveillance.

Types of subsidies, ranging from tax preferences to outright grants, are as varied as the types of goals they are designed to promote, such as employment support or competitiveness in foreign markets. Whatever their type or goal, subsidies generally become so embedded in government and business life that their removal is a rare and wrenching event.

In general terms, the fishing industry may receive several types of subsidy (Porter 1997):

- The first type provides direct income support, either through price supports or through grants to remove vessels temporarily from a fishery. Subsidies of this kind discourage fishermen from leaving a fishery that is economically marginal.
- A second type of subsidy reduces producers' variable costs. These subsidies, including fuel tax exemptions, attract investment into a fleet and enable marginal fishing operations to remain in a fishery.
- A third type, subsidies to the use of capital, attracts investment into a fishery. Subsidies of this type include low-interest loans, loan guarantees that reduce the risk of commercial loans, and tax concessions on investments. Such subsidies play a greater role in marginal fisheries, where commercial banks normally would not provide loans.
- A fourth type of subsidy occurs when governments charge less than they might for exploitation of a public resource. Many governments charge foreign fleets less than they could for access to their fisheries, and most charge domestic fishermen nothing for the fish they catch and sell.
- A final type of subsidy benefits fishing fleets indirectly. This type includes general subsidies to a country's shipbuilding industry, which lower costs for construction of fishing vessels and other vessels. Another example of this type of subsidy is government funding of fish ports and fish-processing facilities.

Some subsidies appear in government budgets and others do not (Milazzo 1998). Of those that do, some appear in the budgets of fishery agencies, where one would expect to find them, and other, often very important, subsidies appear in the budgets of other agencies. Budgeted subsidies also may be divided between those directed at domestic fisheries and those directed at gaining access to fish in the waters of other countries. Subsidies that do not appear in government budgets include subsidized lending and tax preferences.

However categorized, subsidies contribute to oversized fishing fleets and to overfishing. In recent years, governments have launched other subsidy programs in efforts to reduce the size and power of fishing fleets by scrapping vessels or buying back fishing licenses (Porter 1997). Although these programs appear benign, even beneficial, they often have been too small, poorly targeted, and undermined by other subsidies promoting new vessel construction and modernization.

In the late 1980s, growing concern about widespread declines in commercial fish stocks together with increasing size, number, and sophistication of fishing vessels triggered reviews of direct and indirect government economic assistance to fishing fleets in particular. The scale of such economic assistance remained a matter of vague speculation until 1993, when the Food and Agriculture Organization of the United Nations (FAO) published calculations of worldwide fishing subsidies (FAO 1993).

Rather than assembling information on subsidies by individual governments, the FAO study estimated global revenues and costs for fishing and assumed that whatever costs were not covered by fishing revenues might be covered by subsidies. These top-down calculations were disturbing. In 1989, according to the FAO, the world's 3 million fishing vessels incurred total operating costs of $92.2 billion, allocated as follows: $30 billion for routine maintenance; $7 billion for insurance; $18.5 billion for supplies and gear; $14 billion for fuel; and $22.6 billion for labor. Gross revenues amounted to $70 billion, or about $22 billion less than operating costs (FAO 1993).

The FAO regarded this estimated deficit as a minimum because it did not include the cost of depreciation, return on investment, and servicing of debt on the fishing vessels themselves (FAO 1993). Using a conservative rate of 10 percent rather than the 17 percent normally allocated for these costs, the FAO estimated that the world's fishing fleets incurred an additional $32 billion in costs.

The FAO speculated that the $22 billion in unmet operating costs might be explained by several things (FAO 1993). First, open access to fish generally encourages greater investment in vessels and greater expenditure for operating costs than is necessary to catch available fish. However rational individual investments may appear, they collectively produce an economically irrational allocation of capital. The FAO estimated that in 1989, fishing fleets were 30 percent larger and more powerful than necessary to catch available fish.

Also, although fishers may go out of business, fishing vessels generally do not, for several reasons. First, because fishing vessels are specialized in their use, they are not attractive to investors in other maritime pursuits. Second, when fishing businesses fail, they often must sell off their vessels at very low prices. These low prices enable someone else—in the same fishery, in a different fishery, in another state, or in another country—to buy a vessel and enter a fishery with lower fixed costs. As a result, the vessel keeps fishing.

The FAO identified another reason for the large operating losses in the world's fishing fleets: the large subsidies provided by many countries (FAO 1993). Deducting estimated operating costs of $10 billion to $13 billion from estimated revenues of $5 billion, the FAO estimated that the Soviet fleet was losing $5 billion to $8 billion each year, excluding capital costs. Most of these costs were absorbed by government through subsidies of various kinds.

Specific estimates for government subsidy programs were scarce. The FAO report mentioned that the European Union increased its subsidies to the fishing industry from $80 million in 1983 to $580 million in 1990 (FAO 1993). About one-fifth of these subsidies, which did not include subsidies from individual governments, were devoted to the construction of new vessels. The FAO also reported that in the late 1980s, the government of Norway provided its fishing industry with roughly $150 million in subsidies.

The FAO calculations have many drawbacks, some of which the FAO report itself identifies (FAO 1993). A chief drawback was the inherent imprecision in estimated revenues and expenses. The estimates of revenues, for instance, were based on statistics on fish landings reported to the FAO by individual governments. The task of collecting such statistics from the fisheries around Indonesia's 13,500 islands suggests one of the limitations on such statistics. Not only do prices vary widely among species, from $100 per ton for species used for fish meal to $20,000 per ton for sashimi-grade bluefin tuna, but they also vary within species, depending on size, condition, season, and other considerations.

Accurate information on costs was at least as difficult to obtain or estimate (FAO 1993). Crews on fishing vessels generally are not paid a wage. Rather, crew members receive shares of the value of the catch, based partly on their positions on the vessel. Informal credit markets, which are the principal source of capital in many parts of the world, obscure the costs of capital.

Recently, S. M. Garcia and C. Newton (1997) calculated that operating costs for the world's fleet were about $91 billion, or $21 billion more than the revenues of $70 billion. Garcia and Newton calculated that the size of the world's fishing fleets would have to decline by 25 percent in order for revenues to match operating costs. To cover total costs of $116 billion, which includes debt servicing and the opportunity cost of capital, the fleet would have to shrink by 53 percent. Alternatively, fleet costs could be met by a 71 percent increase in ex-vessel fish prices or a 43 percent decrease in fishing costs. Ex-vessel price is the price received by a fisher for fish, shellfish, and other aquatic plants and animals.

Despite the imperfections, the 1993 analysis by the FAO moved the discussion of government subsidies into the open. During negotiations at the United Nations that led, in 1995, to an agreement implementing the Law of the Sea regarding so-called straddling stocks and highly migratory stocks of fish, several governments and nongovernmental organizations cited the FAO study and urged action.

More recently, Matteo Milazzo of the National Marine Fisheries Service (NMFS) in the United States prepared a more detailed assessment of fishery subsidies (Milazzo 1998). Rather than inferring levels of subsidies as the 1993 FAO study had done, Milazzo built his estimates from reviews of government subsidy programs of Japan, Norway, the United States, Russia, China, and the European Union.

Milazzo's estimates for total subsidies should be regarded as conservative, for several reasons. For one thing, Milazzo did not analyze subsidies underwritten by state and provincial authorities, although these are thought to be considerable. Similarly, Milazzo's analysis of China's subsidies was hampered by a lack of information. Nonetheless, he concluded that governments annually provide $15 billion to $20 billion in environmentally harmful subsidies to fishing interests, or at least one-fifth of global revenues from fisheries (Milazzo 1998).

Some Examples

Just as there are world leaders in fishing, there are world leaders in fishing subsidies. Despite their lack of economic soundness and the demonstrable damage they have done to both fish populations and fishing fleets, subsidies transform themselves and persist in the face of criticism. One of the clearest examples of this pattern is seen in the European Union (EU).

In 1983, the EU adopted its first program to reduce the size of its members' fleets even as it continued subsidizing fleet construction and modernization at nearly twice the rate at which it was providing incentives for scrapping or laying up vessels (Porter 1997). As a result, the EU fleets continued expanding. In the late 1980s, the EU itself still was providing $64 million each year for vessel construction and modernization.

Pressure to reduce fleet sizes continued to build, however. A special report of independent experts issued in 1990 concluded that key fish populations were dangerously overfished and called for a 40 percent reduction in fishing capacity as a start (Porter 1997). The EU's Multiannual Guidance Programme (MAGP) for 1987–1991 called for a 3 percent overall reduction in gross tonnage, but only two member countries met their reduction targets; the others increased their fleets' size and power (Milazzo 1997). In most cases, governments continued to provide subsidies for vessel construction. As a result, more than a decade of restructuring produced not a smaller fleet but a substantially increased fleet. For example, the United Kingdom's fishing fleet doubled during the 1990s.

Member states rejected later proposals to reduce fleet capacity substantially (Porter 1997). At the same time, the EU allocated $748 million for fleet modernization in the period 1994–1999, nearly four times the rate of vessel subsidies during 1990–1993. For the period 1994–1999, the EU also budgeted $224 million for port facilities, $705 million for processing and marketing, $102 million for production promotion, and $329 million for aquaculture (Milazzo 1998).

Fishery subsidies by the European Union and its member countries took other forms. To relieve pressure on its own fish populations while maintaining its fleets, the EU effectively redeployed many of its fishing vessels to the waters of other countries (Porter 1997). By 1996, the EU had concluded agreements with fourteen African countries providing access to as many as 1,000 European vessels.

To gain this access for fishermen, the EU paid African countries $350 million annually in fees and grants (Porter 1997). In western Africa, the EU has concluded access agreements with Cape Verde, Equatorial Guinea, Gabon, Gambia, Guinea and Guinea-Bissau, Ivory Coast, Morocco, São Tomé and Príncipe, and Sierra Leone (Milazzo 1997). In eastern Africa, the EU has agreements with Madagascar, Mozambique, and Tanzania; in the Indian Ocean, it has agreements with the Comoro Islands, Mauritius, and the Seychelles.

EU access payments under these agreements benefit mainly Spanish, Portuguese, and French fleets (Milazzo 1997). According to an analysis by G. Porter (1997), tuna vessel owners paid less than 15 percent of the cost of access to tuna under most of these agreements and the EU paid the balance. In addition, EU vessel owners, particularly tuna vessel owners, paid fees below prevailing rates for their catch, partly because the rates they are charged are below market value and partly because the amount of catch is underestimated or underreported.

The EU has also supported domestic fish prices through minimum price support programs, programs to remove excess supplies from the market, and programs to defray storage costs (Milazzo 1997). Moreover, the EU and member states have committed about $50 million each year for generic marketing of seafood, labeling, and quality enhancement.

The EU's continuing program of subsidies partly reflects the region's long history of fishing. But newer fishing powers have also been using subsidies to fuel growth in their fishing fleets. Since the mid-1980s, the government of China's five-year plans have included ambitious targets for increasing aquaculture production and landings from capture fisheries (Milazzo 1997). For instance, the 1986–1990 plan called for a 30 percent increase over the previous plan's levels. Recently, the government announced its intention to raise production further, from 25 million to 35 million tons. Of this, 14 million tons—far more than any other country's catches—are to come from capture fisheries.

China's fish production goals have only partly to do with feeding an expanding population. Like many other developing countries, China views fisheries as ready means of generating foreign currency exchange. Between 1985 and 1995, the value of China's fish exports rose from just less than $500 million to nearly $2.9 billion (FISHCOMM 1997). This was three times the value of China's fish imports.

To meet its production goals, the government has underwritten a dramatic expansion of the fishing fleet (Milazzo 1997). Just in the period 1986–1990, the government's Agricultural Bank of China loaned the fishing industry an average of $800 million per year. Between 1978 and 1994, the fishing power of China's fleet quadrupled. The most direct form of subsidy to the fishing industry has been support for the 10 percent of the fleet that is operated by the government: roughly sixty enterprises with 3,000 vessels and catches of about 850,000 tons.

These and other investments have led to dramatically increased production from aquaculture and capture fisheries. In 1996, China ranked first in the world, having produced 25 million tons in 1995. Of this, about 10 million tons was from capture fisheries; 8 million tons came from China's exclusive economic zone (EEZ) (Milazzo 1997). Under the United Nations Convention on the Law of the Sea, coastal nations have exclusive rights over living resources within the Exclusive Economic Zone, generally within 200 miles of the shoreline.

The growth of the Chinese fishing fleet had a predictable effect on fish populations. By the 1990s, average yields from China's EEZ had fallen to half of the yields of the 1950s (Milazzo 1997). To take pressure off domestic fisheries, the government has recently emphasized the development of deep-sea fishing and overseas fishery bases to provide marketing and transportation support. In 1993, the government began a three-year program to build a distant-water fleet with $3.9 billion in investment for the Indian Ocean, South Pacific, and southeastern Atlantic (Wildman 1993).

In Canada, fishing subsidies helped to both create and maintain a fishing fleet that depleted populations of Atlantic cod by the late 1980s, causing massive economic dislocation. The subsidies began with the best of intentions. Beginning in the 1960s, the Canadian government undertook to expand and modernize its offshore fishing fleets in the face of growing catches by distant-water trawlers from Europe in particular (Porter 1997). Both federal and provincial governments underwrote the construction of new vessels and the purchase of new gear for old vessels. Whatever was not provided through direct grants was covered by low-interest loans.

Even as the government imposed catch limits, it continued subsidizing vessel construction and modernization with outright grants of as much as 35 percent of costs (Porter 1997). In 1978–1979, this program disbursed $11 million in grants. The fleet of large trawlers grew from 524 to 795 off Canada's Atlantic coast, so by 1989, the fleet had five times the catching power needed to bring in the annual quota. Between 1976 and 1980, the number of inshore vessels doubled, financed largely by a provincial agency whose outstanding loans grew by 350 percent.

When these programs increased fishing effort eighteenfold between 1954 and 1968, groundfish catches rose dramatically and then plummeted, hurtling the fleet toward financial ruin (Porter 1997). By 1994, when the Canadian government ceased making loans for fishing businesses, the northern cod stock had declined from 400,000 tons (in 1990) to 2,700 tons.

Rather than allow the messy process of a fleet shake-out to take place in a part of Canada that had little other employment to offer, the government intervened with direct assistance payments to fishermen to help them survive while fish populations recovered (Schrank 1997). In the two decades before the crisis, federal and provincial governments had already spent nearly $4 billion in the fishery, and about half of that had been unemployment insurance payments to fishermen and fish plant workers.

In 1990, the Canadian government announced formation of the Atlantic Fisheries Adjustment Program (AFAP), which projected spending of $584 million over five years (Schrank 1997). Funds were to be devoted to reducing the number of people fishing, to developing new fisheries and economic activities for fishing communities, and to providing funds for laid-off plant workers, among other things. Although it was only partly funded, the AFAP did produce new jobs, but only 250 of them went to any of the 16,000 displaced fishermen and fish plant workers.

In 1992, Canada's minister of fisheries and oceans announced a moratorium on cod fishing as well as a program of emergency assistance payments to fishermen and fish plant workers (Schrank 1997). By the end of the program, the government had made $484 million in such assistance payments.

Other activities, called collectively the Northern Cod Adjustment and Recovery Program (NCARP), were aimed at helping vessel owners weather the moratorium, reducing the number of fishermen through early retirement payments, and purchasing fishing licenses (Schrank 1997). Like earlier programs, the NCARP initially sought to reduce fishing capacity but ended up as an income support program that encouraged fishermen and others to wait out the moratorium rather than seek other lines of work. By 1993, fewer than a third of fishermen and fish plant workers were enrolled in programs that might lead them out of the fishery. Only 1,436 eligible fishermen had taken early retirement, and only 876 had sold their licenses.

In 1994, as the moratorium on cod fishing continued, another government report concluded that the fishing capacity of the northwestern Atlantic fleet was too large by half (Schrank 1997). Instead of directly promoting a reduction in fleet size, the government announced a new five-year, $1.9 billion program. Income payments were to continue but at reduced levels. Intended originally to cover 26,500 fishermen and fish plant workers, the income program eventually benefited more than 40,000. Another $300 million was to be used for buying back licenses and paying early retirement benefits.

After billions of dollars' worth of income payments and other subsidies, the Canadian cod fleet remains much larger than necessary to catch all that the Atlantic cod stocks can produce over the long haul. And Canadian taxpayers appear no closer to overcoming the social and economic dislocation of the fishery's collapse.

When world landings of fish and shellfish peaked in 1989, at levels just one-fifth of those predicted in 1969 (CMSER 1969), it became apparent that future landings would be limited by a lack of fish rather than a lack of boats. The challenge now is to reduce fleets, yet efforts to do so are relatively small and tentative. Furthermore, governments continue many of the subsidy programs that enabled vessels to continue fishing long after fishing was no longer economically or ecologically viable.

Efforts to remove subsidies face tremendous political opposition, particularly where competing sectors of the food industry continue to receive subsidies. A few countries, such as New Zealand, have managed to end subsidies to the food industry. However, most often, subsidies prove politically difficult to end because the interests that rely on them are powerfully motivated to lobby to retain them.

Note

1. This chapter is based largely on a report to the Pew Charitable Trusts. M.L. Weber, "A Global Assessment of Major Fisheries at Risk, Relevant Management Regimes, and Non-governmental Organizations." Unpublished manuscript prepared for the Pew Charitable Trusts, 1998.

✦ Chapter 5 ✦

Managing Fisheries Rationally: Framework and Tools

As the previous chapters have shown, as a global society we have spent more time building our capacity to catch fish than attempting to limit ourselves. The historical view was that the abundance of marine fish was so vast and the effects of fishing so small that there was no need to limit fishing efforts. Serious management of marine fisheries began only in the mid- to late 1970s.

Worldwide, the period from 1885 to 1950 was one of slowly increasing research in fisheries. Marine fishing regulations were very few. The predominant notion was "freedom of the seas," with a recognition that within their own waters, states exercised control over who fished and how much they caught (Iudicello and Lytle 1994). Since the beginning of the twentieth century, countries have entered into multilateral agreements with the aim of regulating fisheries for species that move through their waters.

As discussed in chapter 1, catches continued to increase, particularly after World War II, with the application of technology. As technology lengthened the distance to which nations could send their fleets in search of fish, the idea of extending jurisdiction beyond the territorial sea gained credence among coastal states concerned about protecting "their" fisheries from distant-water fleets. In the late 1940s and through the 1950s, several countries proclaimed fishery conservation zones in their waters and a number of bilateral and multilateral agreements were concluded to create regional and international fishery management bodies (Burke 1994).

In the 1960s, 1970s, and 1980s, fleets expanded at twice the rate of catches. Management of fisheries evolved as fishery law and biological and economic understanding of fisheries changed. Attempts at widespread international agreement on fishery management were unsuccessful, however, until the 1982 United Nations Conference on the Law of the Sea. The treaty developed at that conference, the United Nations Convention on the Law of the Sea (UNCLOS), sets the overall legal framework for fishery management (Iudicello and Lytle 1994). Most important, UNCLOS recognizes the sovereign rights of coastal states over the living marine resources within 200 nautical miles of their shoreline, the area known as the exclusive economic zone, or EEZ. Coastal states are to use the best scientific information available in adopting conservation and management measures to maintain or restore populations to levels that produce maximum sustainable yield. Foreign fleets may gain access to an EEZ only with permission of the coastal state.

More recently, several years of negotiations sponsored by the United Nations led to adoption of an agreement that establishes a framework for managing and conserving stocks of fish that straddle international boundaries or migrate over long distances. The United Nations Conference on Straddling Fish Stocks and Highly Migratory Fish Stocks, known less formally as the Straddling Stocks Convention, was adopted in 1995 but will not enter into force until thirty nations have ratified it.

To various degrees, coastal states have adopted management regimes to meet the requirement for conserving living marine resources within their EEZs. Some domestic management regimes are quite elaborate, such as that of the United States, in which jurisdiction over specific fisheries may be entirely within state jurisdiction, entirely within federal jurisdiction, or a combination of both. Fisheries management within the European Union is even more complex, since it must reflect the policies and views of sovereign states.

Within this legal framework, fishery management agencies use different management measures to meet various objectives, which typically include maximizing economic benefits and employment and, more recently, reducing the effects of fishing on nontarget species. These measures are of three types: output controls, input controls, and technical measures. All three are explored in detail later in the chapter.

Fishery Management in the United States

During the same era of environmentalism that brought the United States the Clean Air Act (1970), the Clean Water Act (1972), the Endangered Species Act of 1973, and the Marine Mammal Protection Act of 1972, the U.S. Congress enacted a fisheries statute, the Magnuson Fishery Conservation and Management Act (1976), designed to promote and develop fisheries in the United States. By the 1980s, most laws and regulations governing wildlife conservation and marine environmental protection had been revisited and made more stringent, but the emphasis on fisheries remained in development mode, including federal financial assistance to encourage and promote new fisheries and fishing power.

The 1976 enactment of the Magnuson Fishery Conservation and Management Act (FCMA) responded to concerns that foreign fleets were exploiting U.S. fisheries. Although one objective of the act—Americanization of U.S. fisheries—was accomplished, the law's conservation purpose has remained elusive. Increased demand for fish and fish products was joined by failed government policies, poorly implemented management, and lack of enforcement. Continued promotion of fishery development in the absence of meaningful access controls substituted domestic overfishing for foreign overfishing (Iudicello 1996).

It was not until the early 1990s that a constituency for fish conservation emerged in the United States. The collective voice was made up of environmentalists, who saw the connections between fishing operations and protection of endangered and threatened marine mammals, and scientists and managers, who predicted and tried to prevent such calamities as the collapse of cod, flounder, and haddock fisheries on Georges Bank, off New England. This event finally served as the wake-up call to all who pressed for reform in fishery management (Iudicello, Burns, and Oliver 1996). More than any other event, the collapse of Atlantic groundfish drove home the notion that the ocean was not limitless and could not infinitely absorb increased technology and effort in the race for fish. Some of the changes that emerged in the most recent reform of U.S. law are explored in more detail in chapter 7.

Resource Management from the Economic Perspective

Now that we have described the legal framework within which governments have granted fishery managers authority to regulate the use of a public resource, we will examine some of the tools they use to do so and the expected economic outcomes. In the previous chapters, we attempted to illustrate that the fishing industry does not work the way most productive industries work. Even so, market efficiency and productive efficiency are relevant concepts in understanding what fuels the race to fish. This chapter illustrates why, when the race to fish is on, government regulations do little more than slow the descent to overfishing and disaster.

To review, the industry will expend more effort than necessary to supply fish as long as there are profits to be made from fishing. These profits, or rents, attract more fishing effort as long as there are earnings to be had from the fishery—even though before long, greater catch for one fisherman means less catch for another. Although there is some level of effort that will maximize the industry's profits from fishing, total fishing effort in an open access fishery will tend toward a point at which the rents are dissipated. Effort levels increase as existing fishermen increase investment or as new entrants come into the fishery seeking revenue, until the rents are dissipated by investment and new entrants.

Possible Policy Approaches—What Can Be Expected?

We have already traced government responses to overfishing from the standpoint of legal regimes. Within the framework created by laws and treaties, managers have a number of tools and tactics they can use to reduce overfishing. They can limit inputs, they can restrict outputs, and they can impose technical restrictions. These tools and tactics can take any or all of the following forms:

1. Input controls
   a. Restricting the number of fishermen working in a fishery
   b. Restricting the amount of gear or the size of vessels
   c. Limiting the period of time over which fishermen are allowed to catch fish, in order to spread reductions equally among them but not reduce the number of participants
   d. Restricting the technology that fishermen are allowed to use
2. Output controls
   a. Imposing landing quotas to limit the total quantity of fish that can be caught
   b. Imposing catch limits per vessel
   c. Imposing size or sex restrictions on the type of fish landed
   d. Imposing technical restrictions such as times and areas in which fish can be caught

Yet another possibility is to tax away the profits that motivate overfishing.

We will examine each of these approaches from the economic perspective and provide examples of how the policies have played out in fisheries.

Limiting Entry into the Fishery

One strategy for controlling fishermen by enforcing rules that restrict fishing effort is to limit the number of participants in the fishery. License limitation has been in place in numerous fisheries around the world since at least the mid-1970s. Also called "limited entry," license limitation programs require anyone fishing in a fishery to possess a license, and the regulating agency authorizes a specified number of licenses. Licenses are generally renewable periodically, may be transferable or not, and may be issued to the vessel, the operator, the crew, or some combination thereof.

At the outset, one would expect to reduce catch by limiting the number of people who can catch fish, that is, issuing licenses to a number smaller than those already fishing. The reduction in total effort would allow the resource to grow, eventually increasing landings. If no other measures are imposed, those remaining in the fishery would realize higher profits. Over time, however, the remaining participants would increase their effort. They might invest in more powerful engines to give them more time at sea or use larger boats with bigger holds. If the only input control is on the number of fishers, the fishers will increase their efforts by substituting inputs that are not restricted until the market returns to the inefficient side of the intersection of total costs and total revenues. This additional investment, called "capital stuffing," is a common experience where license holders find other means to increase fishing power when the licensing program does not otherwise restrict catch or capacity.

An example of this can be seen in the British Columbia salmon fishery. There, managers reduced the number of licensed vessels by 20 percent, but these vessels were then replaced by bigger and better vessels that actually increased the capacity of the fleet. In contrast, in the limited entry program in Alaska, the license limitation includes a limit on units of gear as well as licenses and is linked to a stringent program limiting catch, times, and areas (see chapter 6).

There might be some point at which, if few enough boats are allowed to work in a fishery, some diseconomies of size would eventually keep the remaining participants short of the open access equilibrium.[1] But if the remaining fishermen are allowed to increase individual effort and change their technology, it would be extremely difficult for fishery managers to accurately equate the optimal number of fishermen with the optimal catch. Because the individual skipper can substitute one input for another—for instance, choose to stay on the fishing grounds longer—even if managers applied gear restrictions along with entry restrictions, fishermen would exaggerate other aspects of their capacity in order to overcome the restrictions. Thus, a cycle of regulation, capacity enhancements, and more regulation is set in motion.

Another drawback to license limitation is that imposition of the program in fisheries where effort already exceeds what is necessary to catch the available fish does little to reverse the effects of excess capacity (Gimbel 1994).

Managers must address fairness issues in implementing license limitation, especially if the license limitation results in a fishery unchanged in stock abundance compared with the situation under open access, just with fewer participants. On the other hand, a licensing program that creates a long-term mechanism for reducing fishing effort by grandfathering in current producers but limiting continuation of their licenses after they leave the fishery might be politically more attractive. Fishery managers could then control the number of participating fishermen by means of attrition and controlled replacement. The Alaska program demonstrates how such an approach can work, but at the expense of tremendous time and cost in adjudicating evidentiary and fairness issues to make the initial determination of who's in and who's out.

In the grandfathering scheme a control date or dates is combined with history of an individual's level of participation in selected years (e.g., a minimum amount of landings is required to demonstrate participation in the fishery). Under a lottery alternative, both effective, full-time skippers and others who might not fish as their principal employment would have an equal chance to win a license. Under an auction, the license would go to the highest bidder. In either event, there would be fewer individual fishermen under a restricted entry scheme than under most other policies because limiting effort by limiting the number of fishermen provides a possibility for larger-scale operations. This outcome is also favored with an auction because the auction itself favors larger operators with enough cash to bid on the license.

Seasonal Restrictions

Another way to reduce fishing effort is to reduce the duration of fishing time, that is, to shorten the season. If fishermen catch a given number of fish per day and if the model describes the level that can be sustained over time, theoretically one can divide the total allowable catch by the total catch per day for the number of days in the season. This approach has been used extensively by fishery managers.

Assume that each skipper operates during the course of a normal, unregulated fishing year, at a level of effort at which the marginal cost of effort equals the average revenue for that effort. Under open access, new vessels continue to enter the fishery to the point at which the annual total average revenue equals the annual total average cost—the sum of revenue across the seasons above variable costs just equals annual fixed costs. When one or more of the fishing periods is closed, the annual revenue no longer covers all the fixed costs. Fishers can either leave the fishery or use their capital equipment more intensively during the period open to them. This increases the average cost for the individual fishermen as well as the total cost curve for the fishery as a whole.

Reduced seasons have been found to be as difficult to implement as are limitations on the number of fishermen. As discussed earlier, if the number of skippers is reduced, each will expand his efforts until the marginal costs of catching another fish equal the value of that fish in the market; that is, fishermen will continue to fish away the rents. Similarly, if the amount of time during which fish can be caught is reduced, fishermen will pack greater effort into the shorter period of time. Policies that restrict the fishing season induce fishermen to fish around the clock in a manner that is both dangerous to them and inefficient for society. The predictable behavior of fishermen—continuing to fish as long as there are profits to be made—makes it extremely difficult for the manager to return the fishery to some specific desirable level of reduced effort and higher stocks.

The halibut fishery in the Gulf of Alaska was probably the most extreme example of what happens during a compressed season. There, until 1995, thousands of vessels raced to catch halibut during two twenty-four-hour openings each year, endangering lives and flooding the market with a poorly handled, low-value product (see chapter 6). In New England, managers have reduced the number of days at sea to restrict the groundfish fleet in a rebuilding program. This limitation on time spent fishing has been combined with other measures in the absence of a total overall quota (Fordham 1996). However, an international study has found that these kinds of measures tend to increase enforcement costs and do not ensure conservation (OECD 1997).

Technology and Gear Restrictions

Managers can also control catch by limiting technology—in effect, by not allowing fishers to use gear and vessels they would otherwise find to be the most profitable. This policy has been applied in a number of cases, such as the limitation on hand tonging of oysters in Chesapeake Bay, requirements for hand-pulled set nets in specific salmon fisheries in Alaska, and bans on pair trawls in certain groundfish fisheries. These measures increase the cost of producing a unit of return on fishing effort; they reduce the effectiveness of the inputs the individual skipper brings to the fishery.

Such restrictions can be expected to protect stocks from the use of an improved technology, but they do not necessarily achieve a more desirable outcome for fishermen or fish consumers. Gear restrictions do not maximize economic yield. One thing they can be predicted to do is prevent the predictable capital stuffing that occurs when regulations otherwise curb fishing effort, as described earlier.

The economic effect of a technology restriction is shown by graphing total costs and total revenues under the two different technology scenarios. Box 5.1 shows that technology restrictions might prevent greater stock depletion. However, generalizations are difficult because the outcomes depend on the specifics of the fishery, the technology, and other fact-specific data. For instance, data on whether or not the fishery already was overfished would help determine whether moving from the existing outcome to the outcome economically optimal under new technology would result in more or less fish being supplied in any period. One generalization that can be made is that the catch level obtained under restricted technology could be achieved with fewer economic resources if the new technology were allowed and if catch could be restrained to an optimal level.

Technology restrictions tend to increase danger in fishing. However, they are sometimes attractive to fishery managers because preventing change is typically easier than imposing change. They are by definition technologically inefficient, but they can conceivably prevent a reduction in the long-term supply of fish. Moreover, a recent shift toward use of the precautionary approach in fishery management argues for limiting new technology until managers can review scientific information and determine whether the increased technological efficiency will have a detrimental effect on the stock's capacity to produce maximum sustainable yield. For example, language in the Sustainable Fisheries Act of 1996, which amended the FCMA, requires managers to disallow new

Box 5.1. Effects of a Technology Restriction

This example assumes that a technology improvement only reduces costs and does not result in more fish being caught at the same level of effort. This means that the maximum revenue curve stays put and the cost curve pivots downward.

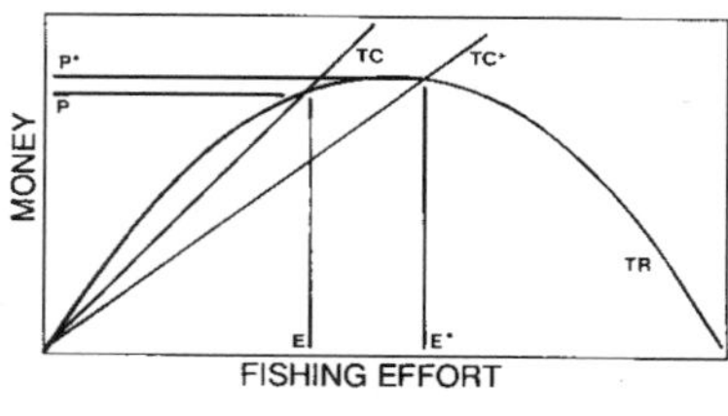

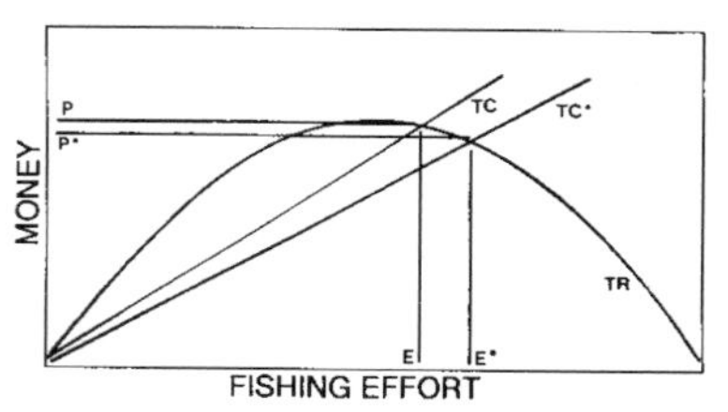

In graph (a), the industry had not been overfishing the fishery in biological terms before the technology improvement. With the technology improvement (represented by line *TC**), the industry suffers under an incentive to overfish stocks because the cost per unit of effort has gone down. However, more fish are delivered to market than were delivered under the old technology. Thus, a technology restriction in this case, although keeping the industry from biological overfishing, also restricts the amount of fish delivered to market.

In graph (b), the industry had already been overfishing before the technology improvement (line *E*). With the improvement, the lower cost of fishing effort encourages fishermen to overfish to an even greater degree (line *E**), thus reducing the quantity of fish delivered to market. If a stock is being overfished, a technology improvement in an open access setting will invariably result in less fish being brought to market over the long run. In this case, a technology restriction results in more fish being supplied over the long run.

gear in a fishery until the proponent of the technology change can prove it will not be detrimental to the stock in terms of habitat or bycatch effects. Bycatch is the unlimited capture of nontarget fish or other marine animals in fishing gear. This shift in the burden of proof runs counter to the traditional regime, wherein the ingenuity and invention of fishers working to improve their returns often leaped ahead of the scientific knowledge or regulatory actions of managers.

Gear restrictions such as limitations on net size and vessel size can also be effective ways to limit catch. Limiting catch by regulating allowable mesh size changes some of the original underlying assumptions about the catch curve in our biomass model, since such a policy would bias the catch (and thereby stocks) toward a certain size class of fish. Whether this increases or decreases productivity of a given fish population would depend on the reproductive biology of the species. It is, in fact, very difficult to compare this policy in general with our economically optimal outcome without first specifying its biological outcome. Further, the ingenuity of fishers looking to maximize profit—even in the face of such controls over their choice of inputs—can be expected to produce new, as yet uncontrolled technology. For example, the invention of "twisted twine," a net-tying technique that creates shrinking mesh, came about in response to an increase in mesh size mandated for conservation purposes in the Atlantic Ocean.

Rather than hypothesize various biological outcomes, one can assume that although restrictions on mesh size might prevent further degradation of a fish stock and might even improve stock levels, there is little reason to expect this policy to move a fishery toward the economically efficient level of either effort or catch. It does not bring about a more efficient level of effort because mesh size restrictions merely limit the available resource to the larger size classes in the fishery. Economic incentives and conditions are not otherwise changed, so one should still expect too much effort to be expended in pursuing this reduced but still profitable resource. Moreover, mesh size restrictions could not be expected to bring about more efficient catch levels because it would be practically impossible to equate maximum mesh size to the stock size for the fishery.

Gear restrictions such as mesh size limits also entail considerable enforcement problems. If a single vessel is fishing for several different species, it might need to carry different nets for its different enterprises. If the regulator allows the vessel to carry different net sizes, it becomes very difficult to enforce restrictions for the regulated fishery.

Other gear restrictions, such as limits on net or line length, have been used for some ocean fisheries. Such restrictions do limit effort somewhat, but they do not provide a very precise instrument for obtaining an optimal catch level, for example, by limiting the numbers of strings of hooks on longlines or the lengths of drift gill nets.

Gear restrictions and similar policies attempt direct control of fishermen's behavior. However, there are less direct economic means of controlling catch levels.

Limiting Catch with Quotas

Quotas can designate an overall catch in an entire fishery, as well as assign portions of total catch among various gear groups, geographic areas of a fishery, or parts of seasons. A quota for a certain portion of the total catch also can be allocated to an individual fishermen. At the most extreme end of the spectrum of quotas would be a moratorium, or ban, on fishing. Under a moratorium, no landings of the restricted fish are allowed and every catcher's quota is zero. This is the case, for example, with jewfish and Nassau grouper, reef species of the South Atlantic Ocean and Caribbean Sea that were nearly extirpated by fishing. In another example, a ban on landings of striped bass in the Atlantic has allowed that species to recover remarkably after severe declines in the 1980s. However, although it cannot be disputed that a total ban on fishing is good for the fish, it causes disruption for both commercial and recreational fishermen as well as consumers.

At the other extreme is management without a quota, using the methods described earlier to limit fishing behavior but not setting a specific limit to overall fishing mortality. The groundfish fishery in New England is still managed in this way. Most fisheries in the United States and elsewhere, however, do have some biologically based overall quota, or total allowable catch (TAC). A TAC sets a maximum level of catch in a fishery for specific species, areas, and time periods (OECD 1997). TACs, which depend on good information about the stock and on effective implementation, are meant to prevent overfishing.

The two fundamental questions attending quotas are (1) how big the total catch should be and (2) how the catch should be allocated. The first question is answered by the graph of catch under the sole owner scenario in box 3.6. If economic efficiency is desired, the optimal level of effort is the one that maximizes the difference between total costs and total revenues for the fishery. The quota, or total allowable catch, should be the one that returns stocks to the left of maximum sustainable yield, where profits are maximized. If biological efficiency is the objective, the optimal catch level is the one associated with maximum sustainable yield because that catch level defines the maximum productivity of the fishery.

Biologists now look for long-term potential yield, the amount of fish or shellfish, in weight, that good management would allow fishermen to catch sustainably. With fishing mortality adjusted to the productivity of the stock, the long-term potential yield would be the most that could be taken from the population without depleting it (Rosenberg 1994).

The second fundamental question in the use of catch quotas is how to allocate them. This question has important ramifications for economic efficiency as well as fairness of the policy. The simplest way to implement a quota would be to announce the quota at the start of the season and then close the fishery after that many fish had been landed, regardless of who has brought them in. In fact, this method is the one most commonly used by fishery managers. But although it is relatively straightforward to implement, it has the same effect on fishing effort as limited seasons. What if a skipper is faced with the potential that the season will be closed (the TAC reached) before he catches enough fish to cover his costs? He will fish harder. TACs, particularly when they limit the catch to an amount below the capacity of the existing fleet, encourage fishermen to use bigger boats and better technology and to work around the clock in order to be the first back to port with a full hold.

In its study of management measures in twenty-five countries, the Organization for Economic Cooperation and Development (OECD) found that management of fisheries by means of TACs generally led to overcapacity in the fishing fleet, shortened seasons, and wide fluctuations in landings. Catching and processing costs rose, and overfishing was not prevented, perhaps because the TACs were set too high or because of poor compliance.

Another approach to quotas, one that takes into account fishermen's economic incentives, is to decide on a total allowable catch and then allocate catch quotas among fishermen. The individual quotas (IQs) limit the catch of individual fishers, and taken together, they equal a TAC (OECD 1997).

Individual quotas can be allocated according to each fisherman's contribution to total catch before the quota comes into effect (based, for instance, on some three-year average for each fisherman and the total catch for that year). Alternatively, equal quotas can be allocated to everyone who wants one; in this case, the size of each individual quota is determined by dividing the TAC by the number of quotas granted. Another method is to auction quotas so that whoever values them most highly wins them. The present system in the United States accomplishes allocation by a process that is principally political—various gear groups or associations of fishermen from particular ports compete by advocating their respective interests before appointed fishery management councils. These regulatory bodies are made up of individuals who represent various sectors in the fishing and processing industry, as well as other experts.

If a quota is binding in that it constrains fishing effort, any allocation system will afflict some fishermen, since reduced effort implies that some fishermen, at least, will be catching fewer fish over the short run. One way to mitigate the imposition on fishermen would be to allocate quotas to fishermen and then promote the free trading of them. In some fisheries, shareholders may trade, rent, lease, or sell their individual quota shares. If quotas are tradable, a market can be expected to develop in which quotas are bought and sold. The quota would represent the right to catch fish, and that right would have value to fishermen. Fishermen who wanted more quotas would be willing to pay a price less than or equal to the discounted returns to catching as many fish annually as allowed by that quota. Obviously, some fishermen would be quota buyers and others would be quota sellers. Thus, along with the boat and gear that anyone leaving the industry would want to sell, fishermen would possess another source of value in their quotas.

If quotas, no matter how they are originally distributed, are tradable, the profits to the fishery will accrue to the holder of the quota. It could be expected that quotas would be gathered by the best fishermen because they would value them most highly and would pay the highest price for them. Less successful fishermen would be less able to absorb the shock of reduced allowable effort, and they would be the sellers of quotas. On the other hand, if quotas are not tradable, a larger number of fishermen might choose to stay in the fishery, since no additional value is attached to getting out. Informal markets for catch quotas might develop that would impose regulatory avoidance costs on operators in the market.

Along with the fishermen, fish consumers also experience a reduction in economic welfare if catch is reduced for a time. In the long run, however, the economic welfare of both consumers and producers will be greater than that attending an overfished open access fishery. But if fishing effort has depleted stocks and effort must be reduced to get stocks back to a more productive level, both consumers and producers as groups will forfeit some welfare in the short run. Economists try to measure these losses and compare them with gains in producer and consumer welfare over the long run.

In cases in which effort is greater than long-term maximum sustainable yield but stocks are not yet depleted, economists predict that the scenario will unfold with industry increasing investment in boats and gear that will ultimately drive stocks downward. Managers will reduce catch and possibly will demand more drastic measures. The further stocks are depleted, the greater will be the dislocation and disruption attending any adjustment to bring fishing effort into line with what is biologically sustainable or economically efficient. If a quota system is introduced before stocks are severely depleted, the pain of adjustment will be less. But if industry and managers cannot agree on a believable production function for the fishery at different stock levels and catch levels, the fishery may have to go "over the edge" before everyone recognizes how much is too much.

Such measures have advantages and disadvantages. The OECD found that IQ programs have generally been effective in controlling exploitation, reducing fleet capacity, and generating profits. On the other hand, they have also proved very difficult in multispecies fisheries. Problems and controversy usually attend the initial allocation of quotas, which is often based on past performance in a fishery. Where rules are not in place, IQs are bought up and concentrated in the hands of a small number of operators.

Taking Away the Profits

Nobody—or everybody—owns the fish. In this situation, in the absence of any rule to the contrary, the rents accrue to whoever goes out and catches the fish. The previous section illustrated why, in an open access setting, this leads to an inefficient result. But what if all the people, a majority of the people, or an authorized decision maker acting through government institutions decides that the rents from a fishery should accrue to fishery management programs or to environmental management in general? This could be done by imposing a landings tax on all fish brought to market.

A tax on landings would act as does an excise tax on oil, coal, or other extracted natural resources and would reduce the amount of effort spent on fishing. As illustrated in box 5.2, the effect of a tax can be viewed as an increase in fishing costs. This increase in costs would lead to a reduction in total effort, though over the short term a perverse outcome could prevail in which fishermen worked harder to catch more fish and each one's net revenue stayed the same. The determining issue is how quickly managers want to restore stocks to their optimal level. In theory, a tax rate could be set that over the long run allows fishermen a reasonable return on capital and that restrains fishing effort to levels at which both producer and consumer welfare are maximized according to market efficiency.

A significant difference between optimally set tradable catch quotas and an optimal landings tax is that with a landings tax, the fishery rents would accrue to the government, whereas with tradable quotas, rents would accrue to owners of the quotas. Such a tax would have the same efficiency as tradable quotas in regard to who stays in the fishery—the productive fishermen. Some fisheries have a landings tax, but it has been used to increase revenue for the taxing entity, not to deter new entrants into the fishery or to force present participants out. Less productive fishermen would not be able to continue fishing if the tax constrained total available revenue and they were competing with more productive fishermen. Unable to compete, these less effective fishermen would be displaced, and only the most efficient fishermen would remain. A tax would also provide an incentive for technology improvements, since they would have the appearance of reducing costs and enlarging profits in the short run. Such technology improvements would require constant adjustments of the landings tax in order to maintain the optimal catch outcome.

Fishermen probably would not want to start paying for something that has always been free, and they might not trust the government to

Box 5.2. Effects of a Tax on Fish Landings

Graphically, a tax on landings (also called a royalty) is either a shift or a pivoting upward of the industry total cost curve. If the tax is a fixed amount per unit landed, the total cost curve shifts upward. The graph shows a sliding-scale tax that increases with increased landings. In either event, along with all the other costs of catching fish, fishermen would, under a landings tax, have to bear the additional cost of catching fish. Thus, they would have no incentive to fish past the socially optimal catch level.

Consider the application of a tax that raises the cost of fishing so that the industry total cost curve *(TC)* intersects the total revenue curve *(TR)* at the economically optimal point *P,* as in the graph. Income will be lost if fishermen fish beyond this level of effort, so they probably will not do it or will not be able to do it for long. The rents would accrue to the government.

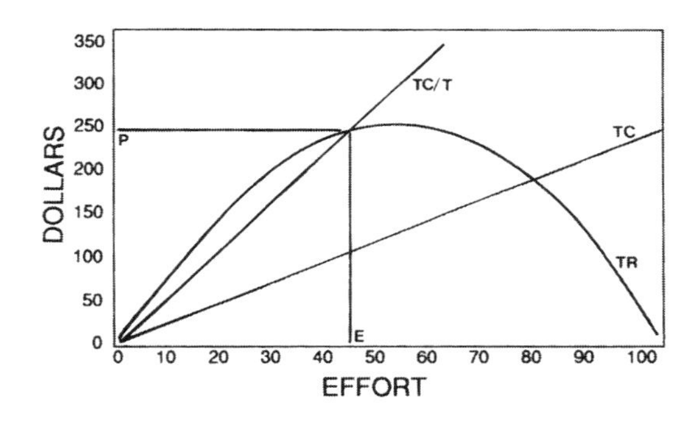

set the tax properly. Policy makers would probably be reluctant to use a tax for the purpose of limiting fishing effort. Moreover, the information necessary to establish the economically optimal long-term total catch quota is considerable: stock assessment, technology, catch levels, market prices, and association of different levels of effort with cost. Theoretically, however, under a landings tax policy, the manager would determine how much tax the industry can bear and still bring the economically optimal amount of fish to market, rather than telling fishermen how many fish each quota holder is allowed to catch.

During most of the 1980s and into the 1990s, voters in the United States elected to restrict the government's share of society's economic resources—a long way of saying "no new taxes." As a result, government has been unable to provide some services, such as fishery information and management, that for a variety of reasons cannot be supplied by private enterprise. In fishery management policy specifically, although a number of states have successfully implemented landings taxes that are used specifically for local fishery management projects and research, in the 1990s the U.S. Congress has rejected proposals by fishery agency officials to allow collection of user fees for federal fisheries. None of this points to success for proponents of a landing tax.

The following chapter provides seven case studies that illustrate how a number of the policies discussed here have been implemented in fisheries around the world, how real skippers behave in fisheries that do not conform to the traditional open access model, and what obstacles and achievements have been recognized.

Note

1. Economies of size occur when cost efficiency is improved by increasing the size of the production unit. At some point, however, greater size starts to reduce efficiency, an effect known as diseconomies of size.

✦ Chapter 6 ✦

Case Studies

In the previous chapters, we described how fish populations respond to fishing mortality, and how fishermen respond to the economic incentive to fish harder in an open access system. We described how managers control fishing mortality through limitations on gear, fishing time, and catch but how in an open access system none of these methods is enough to stop the race for fish. Several options for limiting access were put forth, such as license limitation and individual fishing quotas, along with the idea that such systems change incentives and thus change fishing behavior.

This chapter explores seven fisheries in the United States, Canada, Australia, and New Zealand where one or another limited access regimes has been implemented. The specifics of each fishery allow an examination of how actual fishermen in the real world—not in theory—behaved differently under a limited access system than under open access.

License Limitation in Alaska Salmon Fisheries

In the previous chapter, we noted that a license limitation scheme can be expected initially to reduce productivity, then to slowly reduce the number of fishing units through attrition, and eventually to allow the stock to grow and to increase landings. Used alone, license limitation is expected to lead to investment in more productive gear and vessels and to more time spent fishing because it restricts only one input: the number of vessels. However, if combined with other measures such as catch limits and gear and vessel restrictions, license limitations may adequately control all flexible inputs. The limited entry program in Alaska is an example of the latter: license limitation combined with limitations on catch and restrictions on units and size of gear.

For more than a century, salmon have provided the heart of the North Pacific fishing industry, and for decades salmon have made up the high-dollar fishery in Alaska and the Pacific Northwest. In 1996, this fishery employed nearly 20,000 people in catching and processing and provided $36 million in revenues (NMFS 1997). Historically one of the driving factors in Alaska's push to statehood, the record catches of this fishery today provide the largest source of employment outside government in the forty-ninth state (Browning 1974; NMFS 1996b). After low levels in the 1970s, Alaska salmon catches hit an all-time peak of 196 million fish in 1994, with an estimated value of $427 million (NMFS 1996b).

Each year, five species of Pacific salmon return to the tens of thousands of streams that flow from Alaska's temperate rain forests into the fjords of the state's southeast region and into streams that feed the Chukchi Sea above the Arctic Circle. Some salmon spend longer than others in their natal streams before leaving for the ocean to feed and grow to maturity. Spawning returns generally take place from early spring through autumn, varying by species, races among species, and races within individual streams.

The king, or chinook, salmon *(Oncorhynchus tshawytscha)* is the largest, weighing 15–30 pounds on average, with giants weighing more than 100 pounds. King salmon range from mid-California to the Arctic and mature at three to five years of age. The coho, or silver, salmon *(O. kisutch)* ranges from California to the subarctic and is the salmon most valued by sport fishermen. Coho salmon mature at two to four years of age and weigh 4–16 pounds. Sockeye, or red, salmon *(O. nerka)* are the mainstay of the commercial industry. Historically, the largest runs have been in the Fraser River of British Columbia, Canada, and in Alaska's Bristol Bay. Sockeye salmon mature at four years of age and weigh 0.75–7 pounds. Pink salmon, or humpbacks *(O. gorbuscha),* are the most abundant salmon, ranging from Puget Sound to British Columbia and up into southeastern Alaska. They mature by two years of age and weigh 3–6 pounds. Chum, or dog, salmon *(O. keta)* range from Oregon to the Alaskan Arctic, maturing at three to six years of age and weighing 8–12 pounds.

Although salmon have been a stable source of food for Pacific Rim peoples for thousands of years, the commercial exploitation of Pacific salmon in North America was coincident with the West Coast gold rush (Matsen 1994). After a rich beginning on the Sacramento River in California, the rush moved north to Alaska. The first commercial fishery in the North Pacific was associated with the opening of a cannery in Klawock, in southeastern Alaska, in 1878. The Arctic Packing Company opened the first Bristol Bay operation in 1884, and the Alaska Packers Association started its cannery in 1884 in Naknek, where the company still operates (Browning 1974). Square-rigged sailing vessels made the dangerous journey from San Francisco across the Gulf of Alaska to Bristol Bay each year to cash in on the runs of sockeye. These vessels were used until the 1930s (Matsen 1994). The early years were wide open and wild, as described by Robert J. Browning: "The giants of the industry had almost free rein for decades to act their part as they wished. At times, matters came close to armed warfare, piracy was a fact of existence, the big got bigger mostly, and hundreds of salmon canning companies were born and died" (1974, 43).

Salmon traps, set in streams to catch returning spawners, threatened to decimate the runs and were the bane of the fleet. When Alaska gained statehood in 1959, one of the first acts of the new legislature was to ban salmon traps. Purse seines and gill nets were the gear most often used in taking Alaska salmon, and that continues to be the case today, with some use of trolling in southeastern Alaska and of set nets in Cook Inlet.

Salmon management is conducted by the Alaska Board of Fisheries and the Alaska Department of Fish and Game within state waters (out to three nautical miles). The Board of Fisheries is a constitutionally created regulatory body made up of individuals knowledgeable about the fishing industry. They are appointed by the governor for three-year terms and vote on regulations that govern the state's fisheries. The Alaska Department of Fish and Game provides research, management, and enforcement of regulations. The North Pacific Fishery Management Council (NPFMC), one of eight federal fishery management councils created by the Magnuson-Stevens Fishery Conservation and Management Act, promulgates fishery management plans to govern activities in federal waters from 3 to 200 nautical miles offshore. Since salmon feed in the open ocean, international management is handled by the North Pacific Anadromous Fish Commission (NPAFC), successor to the International North Pacific Fisheries Commission, which governed salmon fisheries in international waters from 1957 to 1992. Members of the NPAFC are Canada, Japan, the Russian Federation, and the United States. The governments of the Republic of Korea and the People's Republic of China have attended commission meetings as observers. The North Pacific Anadromous Fish Commission was established by the Convention for the Conservation of Anadromous Stocks in the North Pacific Ocean, signed in Moscow in 1992, and serves as a venue for cooperation in and coordination of enforcement activities and scientific research.

Alaska's salmon catches increased in the 1990s, but catches of Bristol Bay sockeye were 11 million to 12 million short of the forecast in 1997, and the fleet received federal fishery disaster assistance. Fishery managers attribute the high abundance of the 1980s and early 1990s to informed management, pristine habitats, elimination of high seas drift net fisheries, reduction of bycatch, hatchery production, and favorable ocean rearing conditions (NMFS 1996b). The 1997 season's diminished returns have been attributed to problems with ocean survival due to environmental conditions and to suspected interceptions of salmon in the Russian zone (Benton 1997).

From the 1930s to the 1970s, Alaskans and others continued entering the state's salmon fisheries even though catches were declining overall. Traditional management measures such as closures and gear and vessel restrictions were not sufficient to maintain stocks. By the early 1970s, fishermen and politicians were so concerned about conservation that they spearheaded a campaign that convinced voters to approve the limiting of entry into Alaska's commercial fisheries through an amendment to the state constitution. The ballot initiative was followed by action in the legislature to codify a limited entry program.

Legislature Establishes Limited Entry

The Limited Entry Act of 1973 established a three-member commission, the Commercial Fisheries Entry Commission, also known as the Limited Entry Commission, authorized to limit entry into commercial fisheries to "promote the conservation and sustained yield management of those fisheries and the economic health and stability of commercial fishing . . . by regulating and controlling entry into the commercial fisheries in the public interest and without unjust discrimination."[1] It was clear that the economic health of the fleet was commensurate with the sustainability of the resource.[2] Alaska's legislature expressed its intention to give permit holders a stake in the fishery and an incentive to conserve salmon and obey conservation laws as well as its intention to promote aquaculture to rebuild salmon stocks (Twomley 1994). Within two years, all nineteen of Alaska's primary salmon fisheries were under limited entry management. After twenty-five years, more than fifty fisheries were under limitation, with more than 13,000 permits issued.[3]

Members of the Limited Entry Commission are appointed by the governor, and confirmed by the legislature, to two-year terms.[4] Commissioners are to be persons with broad professional experience but without "a vested economic interest in an interim-use permit, entry permit, commercial fishing vessel or gear, or in any fishery resource processing or marketing business."[5] The commission is empowered to establish a moratorium on entry into fisheries, regulate entry into all state commercial fisheries, establish priorities for the fisheries, establish the maximum number of entry permits for each area and gear type, establish qualifications for entrants, and administer the issuing, transfer, and buyback of all permits. The statute enables the commission to collect fees and to administer its programs in accordance with the state's administrative procedures and rules of evidence and due process.[6]

In Alaska, fisheries are designated by place, type of gear, and resource. Salmon fisheries range from troll-caught king and coho salmon in southeastern Alaska to beach-set gill nets for sockeye salmon in Naknek, drift gill nets for sockeye in Bristol Bay, and purse seines for pink salmon in the Aleutian Islands. The Limited Entry Commission therefore designates administrative areas within which entry is regulated and then designates permits by gear type for particular species.[7]

Of note in Alaska's salmon fisheries is that they are managed not by quotas or total allowable catch but by escapement. Since most salmon are caught when they return to spawn and die in their rivers of origin, managers strive to allow a sufficient number of fish upriver to reproduce for maximum sustainability while allowing the fleet to catch all the fish not needed for escapement (Twomley 1994).

Escapement levels are set by fishery biologists with consideration for run timing, size, species, year class, and so forth. The season opens when the required number of fish have passed upstream. Management measures in the fishery are established by the Alaska Board of Fisheries, which sets seasons, designates gear and vessel limits, allocates among gear types and fisheries, and promulgates other management measures.

The purpose of the limited entry program was not only to establish a moratorium on new entrants but also to freeze the number of units of gear in the fishery at the date of qualification. The commission set a maximum number of permits to be issued, based on the largest number of gear units in the fishery in any one of the four qualifying years. The number of applicants has been much greater than the maximum number allowed. The Limited Entry Commission is expected to gradually reduce the number of entry permits (Twomley 1994). This connection between the license limitation and the limitation on number of gear units is a critical element of the program.[8]

The Limited Entry Commission is authorized by statute to establish qualifications for ranking permit applicants by balancing a number of considerations to ascertain the "degree of hardship which they would suffer by exclusion from the fishery."[9] The considerations include economic dependence on the fishery, percentage of income derived from the fishery, reliance on and availability of other occupations, investment in vessels and gear, and past participation in the fishery.[10] This priority process then establishes categories of permittees, with first priority given to applicants who would "suffer significant economic hardship" and second priority given to those who would suffer only "minor economic hardship."[11]

Through a detailed and circumstance-specific ranking and appeal process using these considerations, the Limited Entry Commission adjudicates applications for permits. Commissioners use qualitative considerations in addition to qualification dates. An evidentiary hearing on the record, final review by the commission, and eventual appeal in the courts make the system time-consuming and expensive. Individuals with challenged permit applications can continue to fish as long as an appeal is pending before the commission or a court (Twomley 1994).

Determining a Level of Effort

Alaska's law enables the Limited Entry Commission to determine whether a fishery should be managed under a limited entry regime, whether a moratorium on new entrants should be declared, how many permits should be issued, and even the number of units of gear that should be allowed.

The commission may establish a moratorium on new entrants for fisheries that are experiencing increases in effort or catch levels that exceed maximum sustainability or for which there is insufficient biological information to manage for sustained yield.[12] During the moratorium, the commission must investigate resource information and fishing effort and determine whether and how to establish a maximum number of entry permits. In making the latter determination, the commission is to consider whether the fishery is distressed (the number of units of gear is more than the optimum) or at a level of participation that requires limitation.[13]

As a starting point, the commission uses the highest number of units of gear used in the fishery during any one of the four qualifying years prior to limitation. It then prescribes measures to reduce the total permits to the optimum number over a period of time, which the commission may specify, through transfers, buyouts, or retirement of permits. The buyouts are partially funded by assessments on holders of entry permits, up to a maximum of 7 percent of gross value of annual catch.

Rights and Transferability

In creating its limited entry permit system, the Alaska legislature declared the permits to be use privileges subject to cancellation or modification by the state without compensation.[14] As such, the permanent entry permit represents a quasi property right (Twomley 1994). The permits are inheritable and are transferable for value. The legislature intended to leave redistribution of entry permits to the marketplace, and there is no provision for the state to reissue permits no longer in use (Twomley 1994). The market value, for example, of a drift gill net entry permit for Bristol Bay salmon in 1994 was $202,000. That price dropped to about $120,000 in 1998 because of a number of factors: an unexplained run failure in 1997, a major antitrust suit pending in Alaska state court against processors and buyers, and the lowest prices ever for salmon.[15] The low price is thought to be a result of significant change in worldwide market conditions, which have shifted dramatically in favor of farmed salmon over wild-caught fish (Johnson 1997). The Limited Entry Commission estimates that the overall value of permanent entry permits is $1.2 billion.

If the number of permits outstanding in a fishery is greater than the optimum number set by the commission, transfers may be made only to the commission.[16] In other cases, the law stipulates that transfers be made only to a living individual, not a corporation or partnership, who can demonstrate present ability to participate actively in the fishery. Holders may transfer permits only after sixty days' notice and may rescind offers to transfer during that time. Permits may not be leased, pledged as security, or executed to satisfy a judgment. In limited exceptions, permits may be pledged in state-authorized loan programs.[17] Creditors' rights against licenses are unclear. The Internal Revenue Service has attempted to seize permits and force their sale but had not concluded any such actions as of 1998. Some court decisions indicate that seizure and forced sale of permits to collect child support might be permissible, but commentators say the child support agencies would "rather have fishermen in the water fishing and earning money" than stripped of their license to fish.[18]

Transfers have tended to favor Alaska residents, even though the Limited Entry Act is neutral with respect to residency and more than three-fourths of the permits are held by Alaskans (Twomley 1994). However, some observers are concerned that the transfers would not benefit rural Alaskans if there were a net drain from coastal communities where economic activities other than those related to commercial fishing are limited (Twomley 1994). On the other hand, some coastal communities have gained permits, and several have remained very stable. In an example of enterprise, the Community Development Quota (CDQ) program in Bristol Bay has started a local permit brokerage that assists fishermen faced with claims from the Internal Revenue Service and child support agencies.[19] The CDQ program also helps guarantee loans to bring more permits back to the area in order to maintain the fisheries' economic base.[20]

Successes and Drawbacks of the Program

Commentators believe that Alaska's license limitation system has met its primary objective: limiting growth in participation in salmon fisheries. Prior to limitation, increasing numbers of participants entered the salmon fisheries even in the face of declining catches. That influx was stopped by the program. Because limited entry permits for salmon fishing are directly related to the number of units of gear in use, the program has served not only to limit participants but also to limit overall fishing capacity. This ability to limit gear and capacity is a critical element of the program.[21] The state legislature has even granted the commission additional authority to limit individual permit holders to a specified historical capacity, based on a tiered system that measures capacity at the time of qualification and then classifies similar groups of participants in similar gear categories.[22]

Conservation and management measures adopted by the Board of Fisheries have helped control fishing power through restrictions on gear and vessels. Coupled with biological management measures, these actions have kept Alaska's salmon populations abundant and, with the exception of king salmon, at a level above their long-term potential yield (NMFS 1996b). In addition, the Magnuson-Stevens Fishery Conservation and Management Act and extended jurisdiction for dealing with salmon interception problems on the high seas have aided in conserving the stocks.[23] Managers claim that the limited entry system gives them more information, enhances their ability to plan, and allows them to make in-season decisions more precisely than was possible under open access.[24]

Whether the license limitation would be applicable or effective in other fisheries is less clear. Commentators believe the system would not be nearly as effective in fisheries in which license holders are not sole owners, effort controls are not part of the entry permit, permit holders can change to larger vessels or add units of gear, or vessel owners do not personally operate vessels or participate in the fishery (Twomley 1994).

Lessons from New Zealand

We proposed in chapter 3 that one expected outcome of creating sole owners in rights to fishing is the generation of rents. As we showed in chapters 3 and 5, as fishers cut the costs of effort in order to maximize revenue, not only does the stock benefit, but also the individual is able to capture the return as rent. After a rough start, one of the earliest individual transferable quota (ITQ) programs is finally demonstrating recovery of rents in New Zealand fisheries.

In 1986, New Zealand launched the most ambitious use of market-based principles in the management of marine fisheries anywhere in the world. Its Quota Management System (QMS) ended open access to 85 percent of New Zealand's fisheries and allocated transferable shares in the catch to fishers. As time has passed and defects in the original program have been addressed, the QMS has become increasingly elaborate. Recent incorporation of stronger standards for ensuring the sustainability of fisheries has increased the program's ability to produce conservation as well as economic benefits.

With its declaration of an exclusive economic zone (EEZ) in 1978, New Zealand gained jurisdiction over 1.2 million square nautical miles of water in the South Pacific Ocean (Falloon et al. n.d.). Of the 1,000 or so species of fish in these waters, 70 to 80 are fished commercially. Catches and value are dominated by hoki, orange roughy, snapper, squid, and rock lobster.

Since the 1980s especially, New Zealand has expanded its fisheries and exports dramatically. In 1995, the country's capture fisheries and

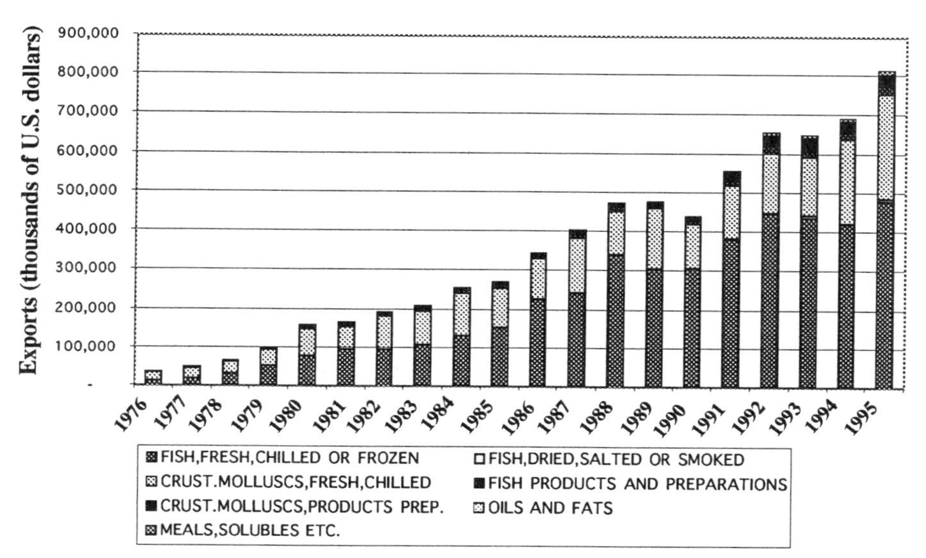

Figure 6.1. Value of Fish Product Exports from New Zealand (*source:* FISHCOMM 1997)

aquaculture operations produced about 654,600 metric tons, valued at $459 million (OECD 1997). Fish and fish products generated U.S.$814 million in export revenues in 1995, up from U.S.$344 million in 1986 (see figure 6.1). The industry employed about 10,000 people and 2,800 vessels.

Until the 1960s, annual landings from New Zealand waters had amounted to less than 50,000 metric tons (OECD 1997). Landings grew rapidly in the 1970s as fleets from Japan, the Soviet Union, Korea, and Taiwan developed fisheries for hoki, orange roughy, oreo dory, southern blue whiting, and squid. Between 1970 and 1977, landings rose tenfold, reaching 500,000 metric tons.

The declaration of an EEZ in 1978 began a transition in New Zealand fisheries that lasted more than a decade. Initially, the offshore catch fell by half as foreign fleets operated under catch limits and other restrictions (OECD 1997). Landings rose again to 400,000 metric tons in 1982 as the domestic fleet entered into joint ventures with foreign fishing concerns and fishing effort grew again. (See, for example, the trends in barracouta landings shown in figure 6.2.)

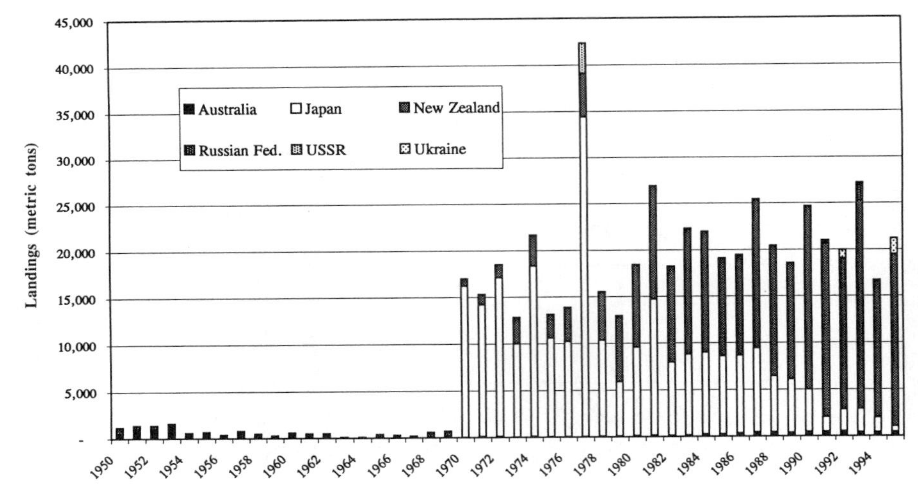

Figure 6.2. Landings of Barracouta from the Southwestern Pacific (*source:* FISHSTAT 1997)

The gradual exclusion of foreign fishing from New Zealand's EEZ fostered heavy investment in fishing vessels, leading to serious overcapitalization in the fleet (OECD 1997). Traditional fishery management methods, such as gear restrictions and closed areas and seasons, failed to halt overexploitation of fish stocks, especially in nearshore areas (Clement & Associates 1997b).

In the early 1980s, management of New Zealand's fisheries began evolving rapidly. In 1982, the government suspended issuance of fishing permits in order to stem the flow of investment into already overcapitalized fisheries. The Fisheries Act 1983 consolidated government fishery management and mandated the development and implementation of fishery management plans (OECD 1997). In 1983, so-called deepwater enterprise allocations were given to nine companies for the seven main deepwater species—hake, hoki, ling, oreo, orange roughy, squid, and silver warehou (Clement & Associates 1997b; Sissenwine and Mace 1992).[25]

In the mid-1980s, concern over the decline of major fisheries as well as general government economic policy favoring privatization, removal of subsidies, and recovery of costs triggered a dramatic change in the way New Zealand managed its fisheries (Dewees 1996).[26] In October 1986, the government adopted the Quota Management System (QMS) for major stocks in New Zealand waters (Clement & Associates 1997b). Principal objectives of this new system of private property rights were the following:

- To rebuild inshore fish stocks
- To ensure that catches were limited to sustainable levels
- To ensure that catches were harvested efficiently, with maximum benefit to industry and to New Zealand
- To allocate catch entitlements equitably on the basis of individual permit holders' commitment to the fishery
- To manage the fisheries so that industry retains maximum security of access and flexibility of harvesting

The QMS process began with allocation of shares of catch to permit holders (Boyd and Dewees 1992). For the deepwater fisheries, such as orange roughy, deepwater enterprise allocations were converted directly into ITQs. For the inshore fisheries, initial grants of quota shares to individual permit holders were based on the best two of three fishing years from 1981–1982 to 1983–1984 (Sissenwine and Mace 1992). As a result, the initial allocations, which were made not as a percentage share of an overall quota but as a fixed amount, probably exceeded the maximum annual catch during the base period. Furthermore, fishers' appeals of initial allocations led to further increases in total allocated quota (Major 1994).

At the same time the government was allocating fixed shares to fishers, it determined a total allowable commercial catch (TACC) based on estimates of maximum sustainable yield and adjusted for recreational and other noncommercial interests in the fishery as well as any other relevant environmental, social, cultural, or economic factors (Sissenwine and Mace 1992). A lack of key information prevented meaningful assessments of many fish stocks in the QMS. Indeed, many TACCs were equated to recent landings, which had been at historical highs. As a result, the TACCs probably were overly optimistic.

Initially, shares or individual transferable quotas (ITQs), which established a right to catch a specified portion of a quota species in a particular area and year, were issued to permit holders in fixed weights and were valid in perpetuity (Sissenwine and Mace 1992). Owners could catch, sell, or lease their quotas.

Fishers who exceeded their individual quotas could purchase or lease additional quotas, could deduct the overage (as much as 10 percent of their total quota) against the next year's quota, or could pay a fee based on the "deemed value" of the catch (Clement & Associates 1997b). In 1997, fees for fresh fish and shellfish ranged from NZ$0.09 per kilogram of jack mackerel to NZ$40.00 per kilogram of spiny rock lobster. Fishers could also enter into an agreement with a shareholder to catch fish under the shareholder's quota.

Fishers who did not use all their quotas could carry forward 10 percent of the quota to the next year. Enforcement of these quotas relied largely on auditing of paper trails between the fisher and the fish buyer. Serious violations could lead to forfeiture of quotas and property and to exclusion from the fishing industry.

At the time, the government operated on the belief that fixed amount ITQs would provide more certainty to industry (Boyd and Dewees 1992). The government planned to buy shares if TACCs fell so that total outstanding shares would not exceed TACCs. The government was to generate revenues for such purchases and for other purposes from industry levies and from the sale of shares the government would grant itself when TACCs exceeded existing shares.[27]

Soon after the QMS was established, the government had to intervene and buy shares because total shares exceeded TACCs in a number of fisheries (Sissenwine and Mace 1992). In all, the government spent NZ$45 million in buying back 15,800 metric tons of shares. Since the number of shares offered by willing sellers fell short of the amount needed to bring total shares in line with TACCs, the government had to reduce shares further by administrative action. The overall reductions in allocation amounted to about 24 percent in twenty-one fisheries.

As biologists learned more about orange roughy and recommended reduced TACCs, the government found itself in the position of having to reduce ITQs even more dramatically (Sissenwine and Mace 1992).[28] In order to reduce the quota shares to levels recommended by biologists, the government may have had to spend more than NZ$100 million. Rather than buying back quotas, the government changed the value of quota shares from fixed amounts to proportional shares of the TACC. To obtain industry support, the government agreed to freeze levies and to redistribute funds raised from the levies to compensate the industry for the reductions.

The size of potential compensation claims associated with decreasing quotas probably discouraged the government from reducing TACCs as much as necessary for orange roughy (Monk and Hewison 1994). After a general agreement that the TACC should be reduced at a rate of 5,000 metric tons per year to an estimated sustainable level of 7,000–9,000 metric tons, the government delayed implementation of the scheduled reductions. The discovery of unexploited populations of roughy temporarily diverted fishing effort away from heavily fished areas.

Early on, litigation established that the QMS effectively disenfranchised the Maori population (Major 1994).[29] The government then adopted the Maori Fisheries Act of 1989 and the Treaty of Waitangi (Fisheries Claims) Settlement Act of 1992, under which the government allocated quota shares to Maori communities. The communities were to receive 10 percent of all existing ITQs, 20 percent of the TAC for any new QMS fish stock, and funding for purchase of 50 percent of the Sealord Group Ltd., the major fish company in New Zealand (Clement & Associates 1997b).

The QMS also required quota shareholders to contribute to the costs of managing and administering New Zealand's fisheries (Clement & Associates 1997b). In the first three years of the program, the government collected NZ$60 million in levies and another NZ$84.2 million from the sale or lease of quotas (Sissenwine and Mace 1992). Initially, the government aimed at increasing resource rentals until the fair market value of quotas approached zero, but industry argued that such an approach did not recognize the risk inherent in fishing. M. P. Sissenwine and P. M. Mace (1992) observed that market price for quotas probably does not measure resource rent because it varies widely and independently of the stocks' productivity and includes extraneous concerns of buyers and sellers, such as the inclusion of other assets in the price.

From the beginning, the government has sought to recover costs and the productive value of fish stocks from shareholders. Initially, the government assessed shareholders a resource rental of NZ$3 per ton (Dewees 1996). Later, rental rates were raised and set according to species; then, in 1994, resource rentals were ended and the government began aiming at recovering the costs of research, management, and conservation.

Today, most funding comes from monthly levies on quota owners and on fishers in non-ITQ fisheries (Clement & Associates 1997b). As an example, in 1997, a fisher holding quota shares for hoki in area 1 paid an ITQ levy of NZ$19.08 and a conservation levy of NZ$0.048 per metric ton of quota share. Depending on the area, a quota owner paid between NZ$57.36 and NZ$146.04 per metric ton of orange roughy quota. In 1994–1995, resource rentals generated NZ$37 million, which covered roughly 80 percent of the cost of management and enforcement in that year (Dewees 1996).

By 1996, the QMS applied to 33 species or species groups in ten fishery management areas (Clement & Associates 1997b). In all, total allowable commercial catch (TACC) levels were set for 185 fisheries. More than 40 other species or species groups are managed outside the QMS, although the government intends to include these under quota management eventually. In the 1995–1996 season, fisheries managed under quota shares accounted for 469,424 metric tons of the total catch of 533,074 metric tons. The leading species under quota shares were hoki *(Macruronus novaezelandiae),* squids (Ommastrephidae), barracouta *(Thyrsites atun),* oreos (Oreosomatidae), ling *(Genypterus blacodes),* and orange roughy *(Hoplostethus atlanticus).*

Effects on the Industry

The introduction of the QMS did not greatly reduce other management restrictions on fishers, such as minimum fish size, closed areas, and closed seasons (Sissenwine and Mace 1992). In some instances, these measures were aimed at conserving fish stocks, and in other instances they were meant to reduce conflict among fishing fleets.

From the beginning, the QMS placed caps on ownership of shares ranging from 10 percent to 35 percent, depending on species. By 1989, the share of catch landed by top companies grew from about two-thirds to 82 percent (Dewees 1996). Sealord Products and Sealord Suisan Ltd. have dominated quota share holdings in a number of major fisheries. In 1991, Sealord Suisan held 31 percent of the TAC for hoki and Sealord Products held 27 percent of the TAC for orange roughy (Falloon et al. n.d.). Since 1989, several large companies have withdrawn from fishing and their shares have flowed to other companies.

Between the 1986–1987 and 1994–1995 seasons, the number of quota owners increased from 1,356 to 1,733 as additional fisheries were included in the QMS (Dewees 1996).[30] The addition of fisheries masked a decline in the number of quota owners in the fisheries that initially came under the QMS in 1986–1987. Of the fishers interviewed by C. M. Dewees in 1987, 23 percent had sold out when the government was buying back shares (Boyd and Dewees 1992).

In an early survey of fishers, Dewees found that most had changed their fishing operations dramatically after the ITQ program began (Boyd and Dewees 1992). Besides spreading fishing out over the year—in some cases to secure the best price possible—many said that they focused on quality rather than quantity.

Early on, high-grading of catch—the practice of discarding fish in order to retain more valuable fish—was a problem in some fisheries under QMS (Boyd and Dewees 1992). This was particularly true of the fishery for snapper *(Chrysophrys auratus)* because of the higher price paid for snapper that met the standards of the lucrative *iki jime* (method of killing fish) market in Japan. Fish that did not meet this standard fetched U.S.$3 per kilogram less. In order to fetch the highest return for each pound of fish caught under an ITQ, fishers discarded fish that did not meet the standard. Only vigorous enforcement brought an end to high-grading in the fishery. It may be, however, that below-standard snapper are entering a local black market.

There is little evidence of whether the QMS reduced the level of capitalization in New Zealand fisheries. New Zealand fleets were overcapitalized by an estimated 20–30 percent, a relatively moderate level compared with that of many U.S. fleets (Lindner, Campbell, and Bevin 1992; Sissenwine and Mace 1992). If this is accurate, there was relatively little opportunity for fleets to consolidate under QMS.

Economic theory predicts that closing access to fisheries, as under QMS, will reduce costs in the fishery and the economy and will generate resource rents for shareholders and government (Lindner, Campbell, and Bevin 1992).[31] The prices paid for purchase or rental of quotas should serve as a basis for determining resource rents and thereby for setting government levies on catch to recover some or all of the rents.

An early analysis failed to calculate precisely resource rents generated in the first years of QMS (Lindner, Campbell, and Bevin 1992). This was partly due to the fact that markets in quota shares do not conform to the economic ideal. Among other things, the prices recorded on registration forms often are fictitious, and quota trades often are barter transactions in which no money changes hands. Uncertainties in TACCs, other regulations, and catches from year to year degrade the quality of information needed for a market to be fully competitive. Finally, vertically integrated firms may well pay a premium for quota shares that will enable productive use of otherwise idle processing capacity.

Nonetheless, R. K. Lindner, H. F. Campbell, and G. F. Bevin (1992) ventured rough estimates of the capitalized value of the fishery and other indices. Data for the 1987–1988 fishing year suggested a value of NZ$765 million. For the same year, estimated resource rents for 15 major quota holders who held 90 percent of all quotas amounted to –NZ$22 million.[32] For the nearly 500 smaller shareholders, resource rents were about zero. These losses are consistent with theory predicting short-term losses in the transition from open to closed access.

QMS and Conservation

Debate has raged regarding the effect of the QMS on conservation of fish stocks. Some proponents of the QMS have argued that ITQ programs should end the ruinous race for fish associated with open access fisheries by investing fishers in the status of fish stocks. However, the New Zealand experience suggests that individual fishers continue pursuing individually rational aims that jeopardize long-term sustainability of fish stocks in other ways (Major 1994). Initially, the race for fish on the water is replaced by a race for quota shares before appeal boards. In the first several years of the QMS, 1,400 of 1,800 fishers appealed the government's record of their catch histories in an effort to increase their allocations. In many instances, additional allocations were awarded to individual permit holders.

Nor did the allocation of ITQs always end pressure to increase TACCs (Major 1994; Sissenwine and Mace 1992). As mentioned earlier, the industry successfully pressed to maintain TACCs for orange roughy at levels well above those recommended by biologists.

In at least one other instance, however, the industry argued against raising TACCs as biologists suggested they could be. The industry's position arose largely from marketing concerns that were independent of the QMS. In this instance, research in the early 1990s indicated that the TACC for hoki could be raised from 200,000 metric tons to 300,000 metric tons (OECD 1997). In addition to expressing concern about the effect of such a change on the settled structure of overseas markets, the fishing industry questioned whether processors could handle additional hoki, especially of the smaller sizes on which a larger catch would depend.

Generally, landings have fallen short of TACCs, sometimes significantly. From the beginning, landings of barracouta have reached only 89 percent of the TACC (1986–1987) and have been as low as 58 percent of the TACC (1993–1994) (see figure 6.3). In recent years, landings have exceeded TACCs in only a few cases. In the high-value

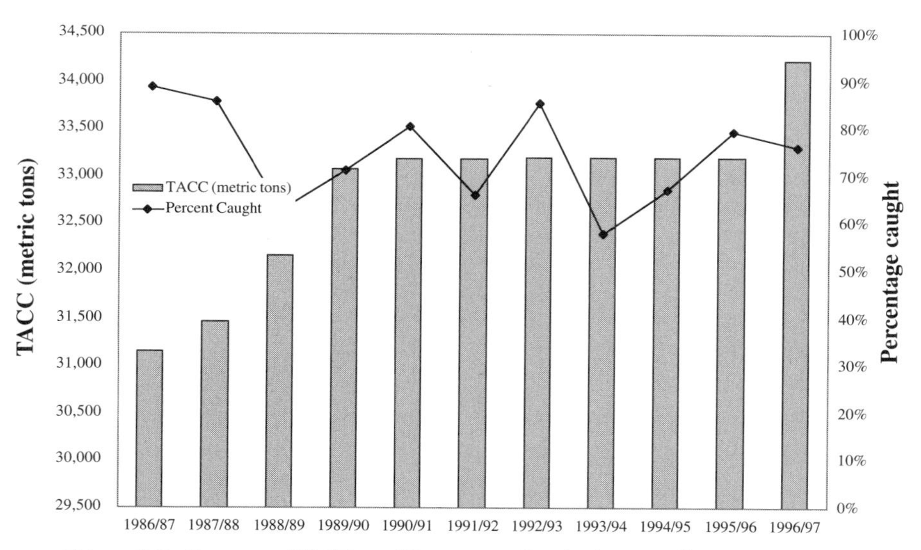

Figure 6.3. Barracouta TACCs and Percentage Caught (*source:* Clement & Associates 1997a)

orange roughy fishery, the TACC was exceeded by 12 percent in 1989–1990, but landings have been at least 3 percent and as much as 22 percent below TACCs since then. These shortfalls are all the more noteworthy since the TACCs were reduced from 42,602 metric tons in 1989–1990 to 21,330 metric tons in 1996–1997.

Sissenwine and Mace (1992) have argued that TACCs have no conservation benefit if they do not restrict catches, as seems to be the case in New Zealand. Either these TACCs are set too high and the stocks are being overfished or the stocks would not be overfished even if there were no TACC.

Bycatch

Bycatch, in trawl fisheries especially, presented other conservation problems for the QMS. Since a trawl fisher may catch five to ten species at a time, anticipating needed quota share for a range of species is a challenge (Boyd and Dewees 1992).

The government sought to address the problem of bycatch in multispecies fisheries in several ways (Sissenwine and Mace 1992). One approach was to assess a tax on fish for which fishers did not hold a quota share. The tax was to be set at a level that would eliminate any incentive for fishers to catch nontarget species.

Among the difficulties with this approach were identification of the appropriate tax level for complexes of fluctuating fish stocks and the ability of vertically integrated fishing companies to absorb even a high tax (Sissenwine and Mace 1992). Provisions allowing fishers to pay a fee for bycatch that exceeded their quotas or to substitute quotas they did hold caused TACs for a number of species to be greatly exceeded (Major 1994). In some fisheries, high levels of bycatch continue; for instance, roughly 80 percent of the catch by weight in the scampi fishery is bycatch (Monk and Hewison 1994).

As is the case under nearly any fishery management system, some fishers in New Zealand have calculated that the gain to be made from illegal fishing exceeds the penalties they might pay and the likelihood that the government will take the time to prosecute them. Misreporting of catches and "quota-busting" in lucrative fisheries such as abalone and snapper have been confirmed (Monk and Hewison 1994). The extent of quota-busting is unknown but is thought to be minor.

Recent Developments

In 1996, the New Zealand Parliament adopted the Fisheries Act 1996, a 390-page document that revises the Fisheries Act 1983 (Chapman Tripp Sheffield Young 1997). The new provisions, which are to be phased in over a three-year period, make adjustments in the QMS without substantially changing its structure.

The Fisheries Act 1996 attempts in a number of ways to address the problems encountered with setting TACCs too high (Chapman Tripp Sheffield Young 1997). First, in taking social and economic concerns into account, the government may only adjust the rate at which a fish stock can be moved toward a size that will produce maximum sustainable yield. Second, TACCs can be increased during a season only for species whose abundance is highly variable. In these cases, TACs are to be set low at the beginning of the year and raised only if there is evidence of high abundance. Finally, the act protects the government from being sued for compensation for the effect of any measure adopted to promote the sustainability of a fishery.

The 1996 act alters several other features of the QMS (Chapman Tripp Sheffield Young 1997). For one thing, owners of quota shares will no longer be able to lease their shares, although they may sell or mortgage them. The act also creates a new element called annual catch entitlements (ACE). An ACE is the actual poundage of catch to which a quota owner is entitled at the beginning of the fishing year. Only 75 percent of a fisher's ACE actually is allocated at the beginning of the year; the balance is to be allocated after about forty-five days into the fishing year. The ACE for one species may be used as cash to pay for the deemed value of bycatch or of catch that exceeds a fisher's ACE. The deemed values charged for catch taken with ACE are meant to remove any economic benefit from catching fish without an ACE.

Summary

What is most distinctive about the New Zealand QMS is that it was comprehensive and systematic (Sissenwine and Mace 1992). Through sweeping legislation, New Zealand adopted a very broad array of measures, from a quota trading system and bycatch controls to buyback schemes and levies for the use of public resources. As the QMS was implemented, economic and conservation problems arose, and these have been addressed somewhat in later legislation.

Causes for concern persist, however. For instance, the funding of government management and research programs primarily from industry levies arguably provides industry with leverage that may be inconsistent with overall social interest in the long-term sustainability of fisheries.

Nova Scotia Groundfish Individual Quota System

We predicted in chapter 5 that individual quota management not only could relieve the pressure of fishing mortality on an overfished stock but also could generate rents where they were otherwise dissipated in an open access regime. Expected consequences of quota management, once the difficulty of the initial allocation is overcome, include a reduction in the number of fishers, particularly if the individual quotas (IQs) are not transferable, because some fishers will not be able to cover their variable costs. We also predicted that since producers will attempt to maximize the value of their quota shares, they will reduce effort, use the least-cost catch methods, and land fish when the price is best. If quotas are transferable, we would expect fishers to trade, with the most productive fishers acquiring the most quotas and increasing their profits. The fishery described in this section illustrates how one group of fishers overcame the hurdle of initial allocation and how some of our predictions played out.

The story of Canadian groundfish management continued to simmer in the headlines in the summer of 1998. The quota for commercial harvest of cod off the southern coast of Newfoundland was doubled, public protests over groundfish management closed a half dozen offices of the Department of Fisheries and Oceans in western Newfoundland, the Canadian cabinet was considering hundreds of millions of dollars' worth of additional aid to the Atlantic groundfish industry, and Spain and Canada squared off before the International Court of Justice over the 1995 boarding of a Spanish fishing vessel in international waters off Canada.

Meanwhile, a quiet success story in the Scotia-Fundy region off Nova Scotia has kept groundfish vessels in business, reduced costs for government managers, and brought about an improvement in stocks of haddock and pollock as well as a slow recovery of cod.

Like most other fishing countries, Canada experienced an expansion of its groundfish fleet in the late 1970s and early 1980s after its declaration of a 200-nautical-mile exclusive economic zone. A second spurt of vessel construction brought on by high fish prices in the United States expanded the inshore groundfish fleet off Nova Scotia even more. Groundfish landings peaked in 1980 at more than 150,000 metric tons. Continued expansion of the fleet and increased exploitation rates did not halt a decline that began just after 1980 and culminated in 1989 with plant closures, layoffs, and hardship among the entire fleet, which was estimated to be four times greater than necessary to catch the quota (O'Boyle, Annand, and Brander 1994).

At the time, individual fishing quotas were not new to Canada's eastern coast, where they had already been implemented in the herring, lobster, inshore crab, and large offshore groundfish fleets (O'Boyle, Annand, and Brander 1994). The government task force that took on the Nova Scotia groundfish crisis recommended that the Department of Fisheries and Oceans (DFO) implement IQs as soon as possible in the mobile gear fleet. Consultations to do so began in early 1991.

The goal of the program was to reduce excess capacity in the inshore mobile gear fleet of vessels less than sixty-five feet in length. The DFO hoped to provide the industry with a means to resolve the problem and avoid continuing disputes about allocation of the shrinking groundfish stocks (O'Boyle, Annand, and Brander 1994).

Recognizing the importance of gaining industry support, the DFO undertook an exhaustive consultative process that included the industry in decision making on issues from the overall shape of the program to the most technical details. An advisory group from the industry decided which stocks to include in the system, determined which fleet sectors would be involved, defined the initial allocation formula, created the appeal process, and deliberated on establishment of a catch monitoring system. The original two subcommittees were eventually replaced with three topical groups: an IQ subcommittee responsible for all regulatory issues for the mobile gear fleet (326 vessels less than sixty-five feet in length), a fixed-gear subcommittee to deal with regulatory issues concerning the fixed-gear sector (about 2,300 hand-liners, long-liners, and gillnetters), and a generalist subcommittee to deal with the concerns of the owners of some fifty mobile gear vessels who decided to remain in the competitive fishery (O'Boyle, Annand, and Brander 1994). The group initially approved IQs for eight stocks of cod, pollock, and haddock. Since then, flounder and redfish also have come under IQs.[33]

By dispensing with several threshold issues, the group was able to focus on the inshore IQ program. First, industry participants agreed not to reopen the existing allocation split between inshore and offshore fleets, leaving the offshore fleet in a competitive fishery. Second, they allowed for the opting out of some inshore mobile gear operators and some fixed-gear operators who wanted to remain in the competitive fishery. The owners who opted out of the IQ program were restricted in amount of quota share and use of gear.

These initial decisions allowed the participants to spend time deciding on allocation arrangements among the vessels in the IQ program. After considering a variety of possible sharing arrangements, industry representatives agreed to base quota shares on a catch history formula, dividing the program into two groups: vessels less than forty-five feet in length (C1) and those forty-five to sixty-five feet in length (C2). They agreed to calculate shares using the average best two of four years' catch history per vessel as a percentage of the 1986–1989 catches for the fishery. Each license was then allocated a quota share based on that percentage (O'Boyle, Annand, and Brander 1994).

Other provisions agreed on were as follows:

- License holders had to fish within two years of the initial allocation or lose it.
- No license holder could retain more than 2 percent of the total of all IQs.
- Allocations were not transferable, although annual nonpermanent exchanges were allowed.
- An appeal process was set up to handle disputes over the initial allocation.

In the initial allocation, licenses were granted to 322 vessels in group C1 and 133 vessels in group C2. Seventy-two appeals were heard, most of them based on extenuating circumstances. By the second year of the program, it was clear that IQ shareholders were transferring quotas, and permanent transfers were authorized.[34]

One feature of the Nova Scotia groundfish IQ program that has proven successful is the catch monitoring system. The DFO and the industry IQ working group hammered out specifications for a private monitoring system to provide 100 percent coverage of landings. The company that was awarded the contract is responsible for authorizing vessels to land, supervising off-loading, and recording and documenting catches. Computerized data are sent electronically to Halifax, where the information is accessible as catch figures by IQ license holder. At the outset of the program, participants set the goal of making the system pay for itself (O'Boyle, Annand, and Brander 1994). Although they did not achieve full cost recovery, the system has provided overall savings for both government and industry. In the first year, the DFO paid the contractor that operates the monitoring system. Now, the industry pays for the monitoring system by collecting fees based on a percentage of the catch.[35] Concerns remain about conflicts of interest in a situation in which the industry is the sole payor of an independent monitoring system, but a DFO program is developing guidelines for dealing with such conflicts and conducting audits of the system. Landing statistics are believed to be much better than those gathered from the previous competitive fishery.[36]

Another concern at the outset was whether the IQ program would lead to bycatch and high-grading. Although the program continues to respond and evolve with regard to sanctions and treatment of bycatch, the original goal of the participants was to stay within the IQ allocation and make efforts to avoid discards. In the event of an overage, sanctions would not apply if the fisherman declares the overage, arranges to cover it with transfers from another IQ shareholder, takes an equivalent amount in value out of his remaining quota in another stock, or surrenders the excess catch to the Crown (O'Boyle, Annand, and Brander 1994).

The program's drafters recommended quota penalties for bycatch and other violations, consisting of both reductions in subsequent years' quotas and removal of fishing privileges for one week to a year, depending on the severity of the violation (O'Boyle, Annand, and Brander 1994). Observers note that the system appears to be working and vessel operators appear to be trying to avoid bycatch by means of operational changes such as fishing times at when nontarget species are not present.[37]

The system has achieved another of the program's goals, fleet downsizing. From more than 300 boats at the outset, the number of vessels participating in the IQ inshore mobile gear groundfish fishery is down to fewer than 100. Some participants moved into the fixed-gear fleet or returned to the competitive fishery, and some left the industry. Although the fixed-gear sector did not adopt individual quotas, it was a limited entry, competitive fishery with a TAC. Initially, the number of participants increased, but they are leaving this sector as well, some transferring to the IQ program.[38]

The pollock and haddock stocks have improved over the years the program has been in place, and the government still maintains a conservative fishing mortality rate and total allowable catch. Recovery of the cod stocks has been slower than predicted or desired, though the program has brought about a reduction in effort and fishing mortality more in line with management targets.[39]

Problems continue in the competitive fishery, which is managed by a community-based system of eight advisory boards situated around the province. Although an IQ program provides structure, managers say that a community-based system designed to do the same thing could also succeed.[40]

Most important, according to observers, is the demonstration that cooperative effort by industry and government can create a structure within which the industry can operate to resolve its own problems. The extensive consultation that has taken place between these two sectors is believed to have played a major role in the program's success.[41]

Surf Clams and Ocean Quahogs

One predicted outcome of individual transferable quotas (ITQs) is that shares will be traded, transferred, and consolidated, thereby reducing effort in the fishery. Other results are a smoothing out and predictability of the market but also a reduction in the labor force employed in the fishery. Although observers will debate the desirability of these outcomes from a social or political standpoint, the experience of the clam fisheries of the mid-Atlantic Ocean demonstrates that the expected economic result ran true to the model.

The fisheries for surf clams and ocean quahogs in federal waters off the mid-Atlantic coast of the United States were the first to be managed by means of controlled access. In October 1990, after more than a decade of public discussion, these fisheries became the first to be managed under a program of individual transferable quotas (ITQs).

For the most part, the ITQ program has behaved as advertised. The amount of capital and labor expended in the fishery has dropped dramatically. The number of fishers and processors has declined less, leading to some concentration of market power. Prices have been more or less stable.

Collection of such long-lived animals as surf clams and ocean quahogs resembles mining more than it does harvest of renewable resources. It should be no surprise, then, that the ITQ program has not averted serial depletion of surf clam and ocean quahog beds off the mid-Atlantic coast. As fishers and processors encounter shortages, the value of ITQs in conserving surf clams and ocean quahogs will be tested.

Surf clams *(Spisula solidissima)* are familiar to consumers as the fried clam strips made famous by the Howard Johnson restaurant chain in the 1960s. Growing to commercial size in six years, surf clams may live as long as thirty-five years (MAFMC 1996). The waters off New Jersey account for nearly half the biomass of surf clams, followed by the Delmarva Peninsula of Delaware, Maryland, and Virginia and Georges Bank off Massachusetts, with about one-quarter of the biomass. In 1995, nearly three-quarters of all surf clam landings came from the waters off New Jersey. In recent years, ex-vessel prices for surf clams have ranged between $11 and $13 per bushel.

Ocean quahogs *(Arctica islandica),* which may live for more than a century, are used principally in soups and chowders (MAFMC 1996). At thirty years of age, they reach a size at which they may be caught by fishers. About half of the biomass of ocean quahogs is located off southern New England and Georges Bank, areas that have not been fished until quite recently as more southern shallow beds have been exhausted. Between 1994 and 1995, landings in New Jersey declined from 3,848 million bushels to 2,177 million bushels and landings from New England grew from 248 million bushels to 1,994 million bushels. Ocean quahogs fetch a much lower ex-vessel price than do surf clams, ranging between $4.00 and $4.50 per bushel.

The higher price of surf clams, which reflects consumer preferences, has resulted in quotas for surf clams regularly being reached (MAFMC 1996). In contrast, the lower price of ocean quahogs has discouraged fishing so that in most years, the annual quota is not even approached. However, recent substitution of ocean quahogs for surf clams in final products has increased fishing effort for ocean quahogs. In 1995, fishers landed 94 percent of the quota of 4.9 million bushels.

The history of the surf clam fisheries falls into three periods. Before 1977, the fisheries were not managed, although the Atlantic States Marine Fisheries Commission recommended a state and federal surf clam fishery management plan in 1976 (Wang and Tang 1993). From 1977 to 1990, the surf clam fishery in federal waters was managed under a limited entry program and the ocean quahog fishery was managed with an annual quota. In the third period, after 1990, the principal surf clam fishery and the ocean quahog fishery in federal waters came under a program whose main feature was allocation of quotas among individual vessel owners.

As in the wreckfish fishery discussed later in this chapter, the fleet fishing for surf clams and ocean quahogs was relatively small even before an ITQ program was introduced in October 1990 (McCay 1994). At that time, 128 vessels reported landings of surf clams; some of these vessels were among 56 that reported landings of ocean quahogs (MAFMC 1996). The number of businesses participating in the fishery actually was less because some owned several vessels (MAFMC 1996). Also, vessels landed their catch in a small number of ports, making monitoring and enforcement easier.

From the beginning of discussions about federal management of the surf clam and ocean quahog fishery in 1977, the Mid-Atlantic Fishery Management Council (MAFMC), established a year earlier by the Magnuson Fishery Conservation and Management Act (FCMA), considered allocating quota shares to vessels (Keifer 1994). Indeed, the notion of allocating quotas in this fishery precedes the FCMA. In 1977, the Northeast Marine Fisheries Board, which had some overlap with the MAFMC, concluded that the fishery should be managed by a "catch-rights" regime in which shares of the quota would be allocated on the basis of past performance.

By November 1977, the MAFMC had received federal approval of a fishery management plan (FMP) for the surf clam and ocean quahog fishery (Keifer 1994). For surf clams, the FMP established an annual quota divided into quarterly quotas, effort limitations such as limits on number and length of trips, and a moratorium on entry of additional vessels into the fishery. For ocean quahogs, the FMP imposed an annual quota, authorized quarterly quotas, and effort limitations.

The first FMP, which became effective in 1979, deferred the allocation of quota shares until later (Keifer 1994). Instead, it imposed a moratorium on entry of additional vessels into the fishery. Fleet owners could not consolidate trips and fishing time for use by one vessel. This moratorium was renewed every year as the MAFMC grappled with instituting a quota allocation scheme in the fishery. Some people argued that allocating quotas to vessels amounted to giving away a public resource. Debate also centered on whether the initial allocation should be based on catch history as established by logbooks or on vessel size.

One effect of the moratorium on entry of new vessels was that owners could not replace unsafe or marginally effective vessels (Raizin 1992). As a result, they attempted to keep these vessels fishing as long as possible. Some engaged in capital stuffing by substantially increasing the width and length of their vessels, installing larger hydraulic pumps, or increasing the size of their crews (Wang and Tang 1993).

This dynamic was further encouraged by later MAFMC decisions. To help establish a fair means of establishing dependence on the fishery as reflected in catch records, a panel proposed a freeze on the size of dredges used to collect surf clams (Keifer 1994). The panel had no sooner released its draft in 1979 than fishers began installing the largest dredges their boats could carry in an effort to build a history of large catches.

When the MAFMC reopened the discussion of allocating shares in 1981, the controversy and confusion deepened (Keifer 1994). Some argued that shares would become concentrated in a few hands and that independent operators would disappear from the fishery (McCay 1994). As it was, one processing company already had a much larger share in the fishery than others because it owned a large fleet of fishing vessels. Some people argued that new fishers should be able to enter the fishery without having to buy a permitted vessel (Keifer 1994). Others argued that crew members should be included in the initial allocations, since they are co-venturers rather than employees under law and custom (McCay 1994). Although they do not contribute capital to the fishing venture, they do share in the risks and returns.[42]

By the late 1980s, management of the surf clam fishery was consuming enormous amounts of time in debate, analysis, and application of ever more restrictive management measures (Keifer 1994). By 1990, the average annual number of fishing hours allowed under the limited entry program had shrunk from 456 hours per vessel to just 154 hours (Wang and Tang 1993).

These developments contributed to the MAFMC's proposal of an individual transferable quota (ITQ) program in July 1988. The proposal included formulas for initial allocations based on fishing history since January 1979. Some fishers insisted that these allocations should include investment as a factor, as measured by size of fishing vessel.

Although vessel owners agreed generally on the benefits of an ITQ program, they could not agree on the specifics (Keifer 1994). No matter which formula was used for the initial allocation, some owners benefited more than others. Lack of agreement over the allocation formula led Congress to ask that the MAFMC delay action while consensus was sought.

After more than a year of review, the MAFMC adopted the ITQ program, and it was implemented in October 1990 (Keifer 1994).[43] With approval of Amendment 8 in 1990, the MAFMC changed the fishery's management objectives and adopted a program of individual transferable quotas (ITQs). The reformulated objectives were as follows:

- Conserve and rebuild Atlantic surf clam and ocean quahog resources by stabilizing annual harvest rates throughout the management unit in a way that minimizes short-term economic dislocation.
- Simplify to the maximum extent the regulatory requirements of clam and quahog management to minimize the government and private cost of administering and complying with regulatory reporting, enforcement, and research requirements of clam and quahog management.

- Provide an opportunity for industry to operate efficiently, in a manner consistent with the conservation of clam and quahog resources, which will bring harvesting capacity into balance with processing and biological capacity and allow industry participants to achieve economic efficiency, including efficient utilization of capital resources.
- Provide a management regime and regulatory framework that is flexible and adaptive to unanticipated short-term events or circumstances and consistent with overall plan objectives and long-term industry planning and investment needs.

In the end, the surf clam and ocean quahog program included no caps on ownership of ITQs (McCay 1994). Also, the program allocated ITQ shares not to vessels but to owners, thereby allowing an owner to consolidate shares from several vessels. In these ways, the surf clam and ocean quahog ITQ program more closely followed the principles of an economics-based, market-driven approach (McCay 1994).

Initial allocations went to owners of 154 surf clam vessels and 117 ocean quahog vessels. In all, there were 67 unique surf clam ITQ owners and 52 unique ocean quahog ITQ owners.[44] Formulas were based on the previous nine years of a vessel's catch history, with the last four years doubled and the lowest two years dropped (Raizin 1992). Different formulas were used in allocating quotas to vessels in the different geographic segments of the fishery. For vessels fishing for ocean quahogs or fishing for surf clams only in New England waters, only catch history was used; for vessels not fishing in New England waters, vessel size was also included as a factor. Effectively, these formulas most benefited vessels that performed well after the moratorium on vessel entry was imposed.

To reduce costs, the plan relied on tracking of landings based on reports from both fishers and shoreside processors (Raizin 1992). Both vessels landing surf clams or quahogs and the land-based processors purchasing them were to complete logbooks identifying number of landings and various tracking information, such as numbered tags on the cages in which clams or quahogs were transferred from vessel to processor.[45]

Owners of shares could use their allocations in any of several ways (Raizin 1992). They could hold the allocation, lease it to others, or sell it. Because the entire surf clam quota could be caught by as few as 13 vessels operating full-time, vessel owners soon consolidated their quota shares and retired old vessels (Wang and Tang 1993). In the first year, the number of active vessels in the fishery fell from 133 to 76, and the full-time fleet shrank by 64 percent, dropping from 77 vessels to 28 vessels (Wang and Tang 1993). Compared with the peak of the fleet in 1986, the surf clam fleet had shrunk by 8,017 gross registered tons. As predicted, however, the average landings per boat increased, moving from approximately 25,000 bushels in 1990 to nearly 70,000 bushels in 1995 (see figure 6.4).

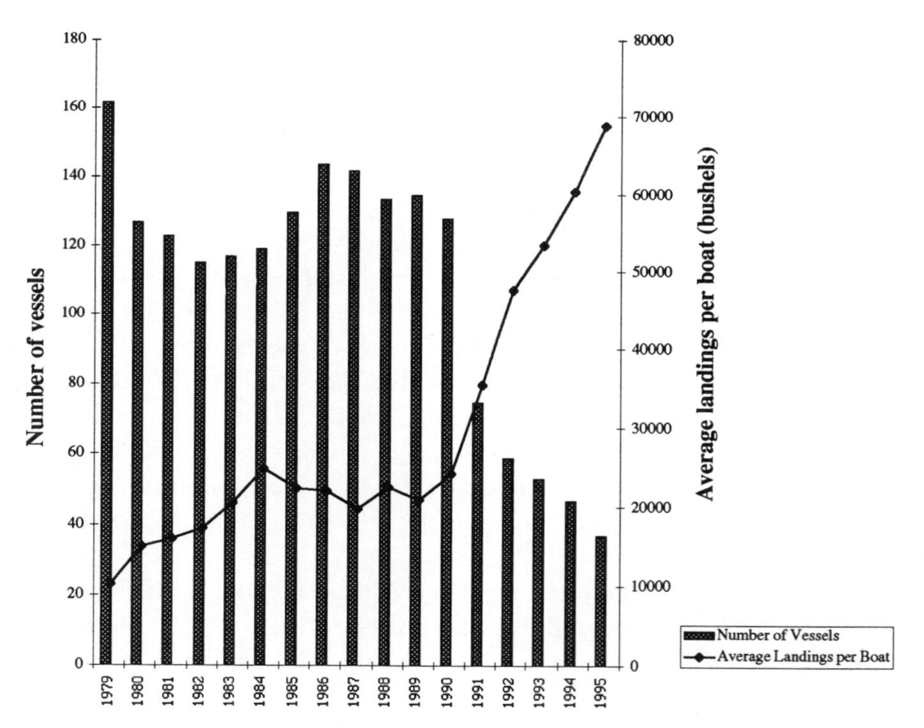

Figure 6.4. The Mid-Atlantic Surf Clam Fishery: Vessel Performance (*source:* Mid-Atlantic Fishery Management Council. 1996. Overview of the Surf Clam and Ocean Quahog Fisheries and Quota Recommendations for 1997 and 1998. MAFMC Dover, Delaware)

The labor force shrank as well under the ITQ system (Wang and Tang 1993). Under the limited entry system, vessel owners could not consolidate fishing time allocated to their vessels. As a result, most crew members were employed part-time until fleet owners began consolidating crews and having them fish with two or three vessels. In the first year of the ITQ program, the labor force fell to 328, the lowest since 1980.

In the view of the MAFMC, the management burden of the fishery declined. After the ITQ program was adopted, discussions were limited largely to the setting of annual quotas.

Conservation and Economic Efficiency

Proponents of the ITQ program in the surf clam and ocean quahog fishery argued that ITQs not only would lead to more efficient use of economic resources but also would promote conservation of living resources (Raizin 1992). Some proponents even argued that ITQs would encourage a conservation ethic among fishers by vesting them in the future productivity of the fishery resources. However, as in the wreckfish fishery, only anecdotal evidence for a conservation ethic is available. This is largely limited to an end to lobbying by fishers and processors for higher quotas or relaxation of effort limitations.

Like any other management program that relies on a total allowable catch, ITQs promote conservation by limiting catch (Raizin 1992). The Mid-Atlantic Fishery Management Council adopted a goal of maintaining surf clams at a level that would permit a take of 1.85 million to 3.4 million bushels per year for at least ten years and that would meet estimated annual demand (MAFMC 1996). From 1990 to 1996, the MAFMC reduced quotas for surf clams from 2.85 million bushels to 2.565 million bushels because of declining catch rates (MAFMC 1996). Between 1987 and 1995, the average number of bushels of surf clams harvested per hour of fishing in the most productive areas, close to shore off New Jersey, declined by nearly half in one area (MAFMC 1996).

As catch rates declined in nearshore areas, fishers moved farther offshore, increasing landings from 275,000 bushels to 1.5 million bushels in 1994 (MAFMC 1996). Landings from these areas declined in 1995 by 17 percent. Future increases in landings will have to come from more southern areas off Maryland, where catch rates reach 200 bushels per hour. In the past, fishing has not been heavy in this area because surf clams there are smaller than in the waters off New Jersey. Distance from shore and the need to test the catch of surf clams for the presence of the PSP toxin, which causes paralytic shellfish poisoning, have made fishing off Georges Bank prohibitively expensive.

For ocean quahogs, the MAFMC had adopted as a goal the maintenance of populations at a level that would enable the taking of 4 million to 6 million bushels per year for the next thirty years and that would meet estimated annual demand (MAFMC 1996). For the 1995 season, the MAFMC reduced quotas from 5.3 million bushels to 4.45 million bushels as the fishery exhausted bed after bed of ocean quahogs. By the late 1980s, when landings off Virginia and Maryland fell from 130–150 bushels per hour to 60–90 bushels per hour, the fleet had moved to waters off New Jersey. After landings off New Jersey peaked in 1991, the fleet moved to waters off Long Island, and by 1995, catch rates there had declined from 146 bushels per hour to 98 bushels per hour.

By 1996, the fleet was working the waters off southern New England, where catch rates of 130–200 bushels per hour prevailed (MAFMC 1996). The fleet was prevented, however, from entering Georges Bank, which had been closed in 1990 because of presence of the PSP toxin.

In 1996, an advisory panel concluded that the slow growth rate of ocean quahogs precluded the fishery from persisting for thirty years unless it could exploit ocean quahogs on Georges Bank and in deep-water areas off southern New England that were beyond the reach of available technology (MAFMC 1996). Indeed, if catches were to remain at recent levels in areas south of New England, ocean quahog stocks would be exhausted in these waters within twelve years.

Evidence of greater economic efficiency in the fishery is clearer. Theoretically, an ITQ program should encourage vessel owners to plan the use of their vessels in the most efficient manner possible and without fear of losing catch to another vessel, as occurs in open access fisheries (Raizin 1992). The decline in the number of vessels and crew members in the fishery implies savings that made capital and labor available for use in other sectors of the U.S. economy (Wang and Tang 1993). These savings came at the cost of some economic and social dislocation that has not been evaluated.

Some of those who received initial allocations of shares sold out, and others consolidated or acquired additional shares. In the first two years of the program, the number of unique owners of ITQs declined (Wang and Tang 1993). Unique owners of surf clam ITQs fell from 67 in October 1990 to 50 in March 1992. The number of unique owners of ocean quahog ITQs fell from 52 to 39. During the same period, the four largest owners of ITQs increased their shares from 51 percent to 58 percent of the total.

After this initial consolidation, the number of share owners remained nearly the same.[46] In 1998, the top 10 firms, of a total of 108 owners of quota shares, held 42 percent of the surf clam shares. Of 65 ocean quahog firms, the top 6 owned 54 percent of the ocean quahog shares.

Vessels remaining in the fishery were more fully utilized after the ITQ program began (Wang and Tang 1993). Since the ITQ program removed limitations on fishing times, vessel operators could fish whenever they wanted and for as long as they wanted. Annual fishing hours rose by 65 percent, to 254 hours, for the surf clam fleet and by 90 percent, to 850 hours, for vessels targeting both surf clams and ocean quahogs.

As with other ITQ programs in the United States, where federal law has until recently prohibited the charging of fees for use of fishery resources, the initial allocation of quota shares involved the transfer of rents, in the form of market value of the shares, to vessel owners in the fishery without compensation to the public (Raizin 1992). On the basis of the prices reportedly paid for purchase or lease of quota shares, early evaluations suggested that the resource value of surf clams and ocean quahogs amounted to $84 million (Wang and Tang 1993).[47] These "assets" generated annual resource rents of $14 million to the owners of quota shares. Surf clams accounted for $57 million of the value and $11.4 million of the rents.

One common theme of debates over proposed ITQ programs is the concentration of shares in the hands of a few businesses. Besides focusing on concerns about equitable distribution of wealth, these debates also touch on the power of firms to influence the price paid to fishers for their catch and the price paid by consumers. Several measures may be used in analyzing concentration in markets (Wang and Tang 1993). One such measure, the Gini coefficient, determines the degree of relative concentration of market shares. This coefficient is zero when firms in a market are equal in size; it approaches 1 as the case of a single buyer or seller in a market is approached. A second measure, the concentration ratio, is the percentage of market shares that a certain number of the largest firms hold. Finally, the Herfindahl index measures competition in the market on the basis of the distribution of shares among all firms. The higher the index, which ranges between zero and 10,000, the less competition there is in the market.

During the decade before institution of the ITQ program, buyers of surf clams and ocean quahogs numbered only twelve to nineteen firms (Wang and Tang 1993). As a result, the Gini coefficient was quite high. Furthermore, the largest buyers bought most of the landings, accounting for 58–74 percent of the surf clam market. Likewise, the Herfindahl index was quite high. Immediately after implementation of the ITQ program, a decline in concentration in the surf clam market moderated. By contrast, after the ITQ program began, growing concentration in the ocean quahog market reversed. In the combined surf clam and ocean quahog market, the number of buyers declined after the ITQ program began and market concentration reached historical levels. The three largest firms controlled 68 percent of the market.

Other changes occurred among processors. For instance, before the ITQ program, some processors sought to gain an advantage by obtaining surf clams and ocean quahogs both from vessels they owned and from independent vessels (Wang and Tang 1993). Other buyers bought only from independent vessels. With the ITQ program, the advantage of owning a vessel vanished and the advantage went to the owner of quota shares. As a result, vertical integration of harvesting and processing activities became irrelevant. Some processors relinquished the operation of vessels and concentrated on acquiring shares.

The influence of this level of concentration on the consumer market is unclear. On the one hand, the price of surf clams declined to its lowest level in 1992, whereas the price of ocean quahogs rose, continuing a trend several years old by the time the ITQ program was in place (Wang and Tang 1993) (see table 6.1). This phenomenon is partly explained by the growing substitution of ocean quahogs for surf clams in traditional markets.

The surf clam and ocean quahog ITQ program has performed

Table 6.1. Ex-vessel Prices for Surf Clams and Ocean Quahogs

Year	*Surf Clams*	*Ocean Quahogs*
1983	85	98
1984	94	98
1985	103	98
1986	104	111
1987	88	106
1988	88	103
1989	88	103
1990	86	111
1991	84	124
1992	83	124
1993	88	128
1994	118	129
1995	118	136

Source: NMFS 1990, 1996a.
Note: 1982 = 100.

more or less as proponents have argued it would. Old, unsafe vessels have departed from the fishery, leaving a fleet that is more fully used. Fishers may spread their activities throughout the year rather than fishing only during narrow regulatory windows. As measured by ownership of shares, influence in the market is only a little more concentrated than it was before the ITQ program.

A comprehensive evaluation of the ITQ program in this fishery will remain unfeasible as long as key socioeconomic information, including employment figures and trading prices for shares, is not collected.

Wreckfish in the South Atlantic

Another predicted outcome of individual quota programs is a slowing of fishing effort in order to produce the quality and size of fish most demanded by the market, accompanied by a commensurate increase in price. The following case illustrates how the market and the ability of quota share owners to respond to it took the pressure off not only the fishermen but also the fish.

The waters off the Atlantic coast of the United States have been heavily fished since at least World War II. By the 1980s, fisheries for the commercially more valuable species had fully developed. Many commercial and recreational fisheries had reached the common paradox of declining catches and rising fishing effort. Generally, new fisheries were limited to species that previously had been discarded but had become attractive in the market (Gauvin, Ward, and Burgess 1994).

In 1987, a fishery targeting a species new to U.S. fishers began to grow rapidly. Fearing overcrowding and overfishing, the South Atlantic Fishery Management Council (SAFMC) moved quickly and adopted the second U.S. management program to rely on individual transferable quotas (ITQs). By most accounts, the program has been a success. Unlike most other such programs, it was adopted early in the development of a fishery.

In the late 1980s, swordfish long-liners began reporting that they were finding thirty-pound wreckfish caught on lost longlines recovered in an area off Georgia called the Charleston Bump (see map 6.1).

Wreckfish *(Polyprion americanus),* which may reach a length of five feet and a weight of 100 pounds, are relatives of the more common striped bass and giant sea bass (McClane 1974). In the western Atlantic, where they are somewhat rare, wreckfish are found at depths of 250 to 400 fathoms on the Blake Plateau, 120 miles southeast of Savannah, Georgia (Gauvin, Ward, and Burgess 1994). Using otoliths as well as length and weight measurements, biologists have calculated the maximum age of wreckfish caught in the fishery off Georgia as thirty-one years (SAFMC 1997).[48]

Wreckfish mature at six to eight years of age and first appear in the U.S. fishery at about eleven years of age (SAFMC 1997). Peak catch rates occur in early summer, when wreckfish gather for spawning. It is unclear whether the fishing grounds off Georgia are the only spawning area for wreckfish in the North Atlantic.

Although wreckfish are new to U.S. fishers, they have been fished since 1960 in the eastern Atlantic, where landings increased during the 1990s from 498 metric tons in 1991 to 1,133 metric tons in 1996 (see

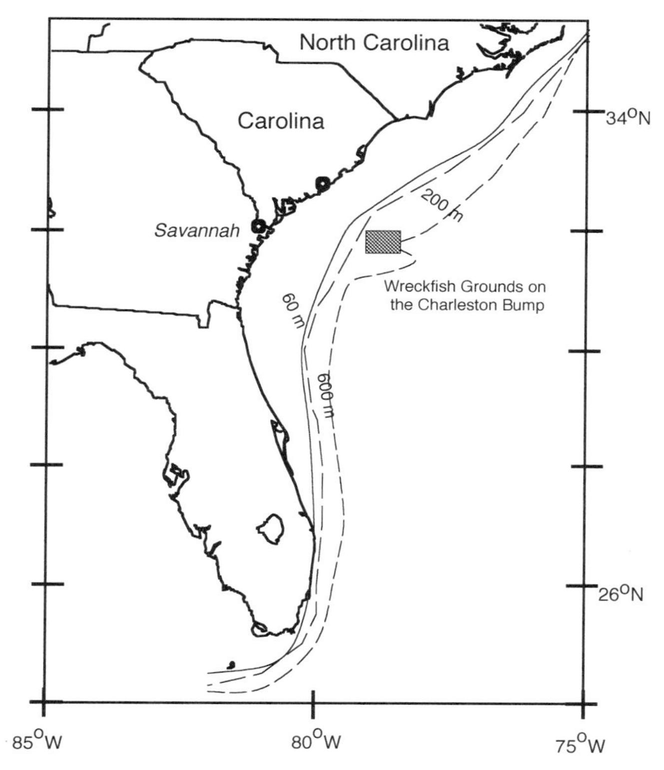

Map 6.1. Location of wreckfishing grounds on the Charleston Bump, a topographic feature on the Blake Plateau, off the southeastern United States.

figure 6.5). In Portugal, wreckfish fetch a higher price than does another demersal species (Gauvin, Ward, and Burgess 1994). In the U.S. market, wreckfish soon became a substitute for grouper, which were becoming scarce. In 1997, the National Marine Fisheries Service classified several species of grouper in the U.S. South Atlantic as overfished (NMFS 1997c).

In 1987, after the inadvertent discovery of wreckfish by longline fishers, two U.S. vessels began fishing for wreckfish using lines suspended from hydraulic reels. In 1988, six vessels fished for wreckfish and landings increased dramatically to 617,662 pounds (SAFMC 1997) (see figure 6.6).

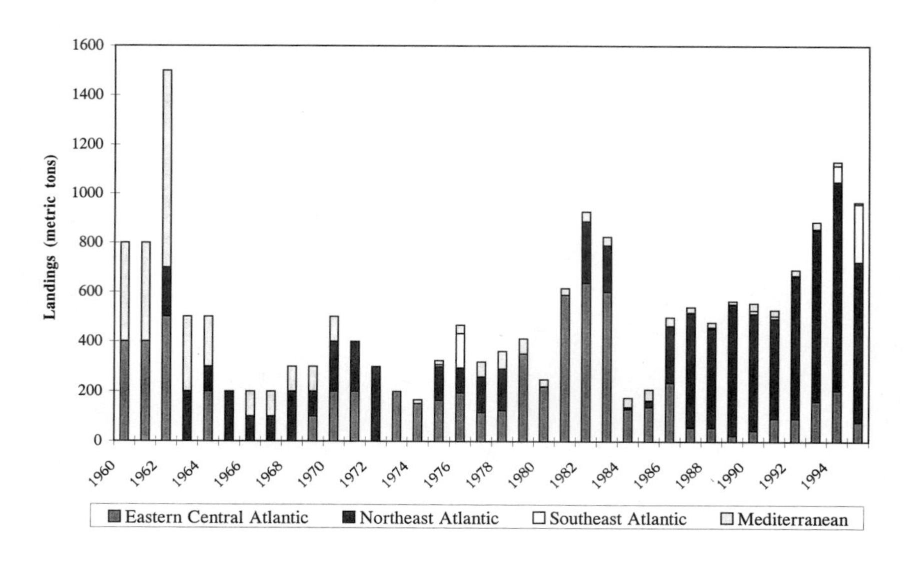

Figure 6.5. Atlantic Landings of Wreckfish (*source:* FISHSTAT 1997)

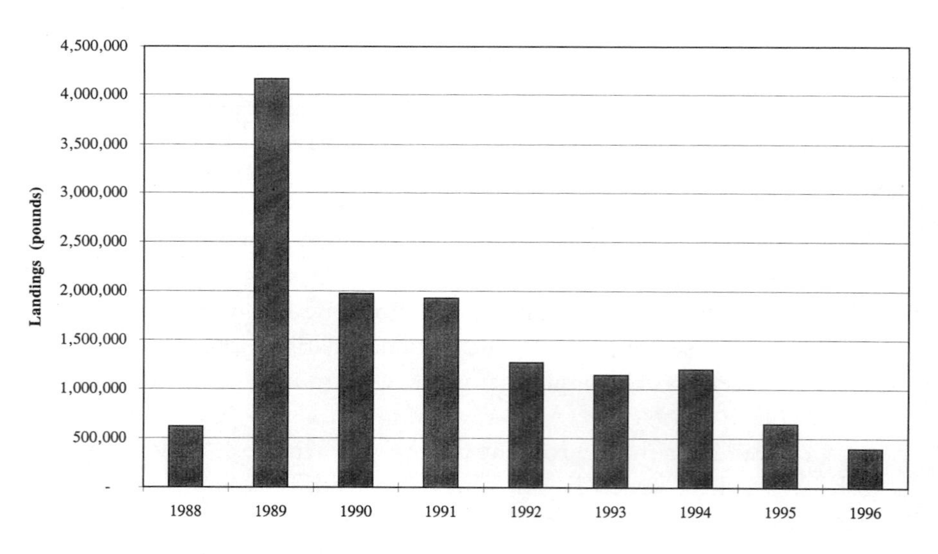

Figure 6.6. Wreckfish Landings in the U.S. South Atlantic (*source:* SAFMC 1997)

Vessels poured into the fishery, attracted by relatively high prices and landings (Gauvin, Ward, and Burgess 1994). At $0.90 to $1.55 per pound, the average catch of 7,000 pounds for a weeklong trip made the fishery one of the most lucrative in the region. Net returns to owners and operators ranged from $35,000 to $97,000 for fishing vessels worth $180,000 to $250,000 (SAFMC 1990). Rigging of a fifty- to sixty-foot vessel cost as little as $15,000.

In 1989, more than forty vessels caught 4.2 million pounds of wreckfish (SAFMC 1997). The South Atlantic Fishery Management Council began imposing controls on the fishery in 1990 (Gauvin, Ward, and Burgess 1994). These included a procedure for setting a total allowable catch (TAC), a control date for establishing eligibility for any limited entry program, limits on catch per trip, and restricted off-loading hours. The SAFMC also changed the fishing year from January–December to April through March, which allowed 4 million pounds to be caught in calendar year 1990 (Gauvin 1994). For the fishing year beginning in April 1990, a TAC of 2 million pounds was set.

One consequence of managing fisheries by TACs is the creation of so-called derby or Olympic fisheries, in which fishers race to capture as much as possible of their quota as quickly as possible. Despite fishing early in the year, when catch rates normally are low, fishers landed the 1990 wreckfish quota in four months (Gauvin, Ward, and Burgess 1994). The race for the fish sent fishers offshore in bad weather and generated conflicts among them in the small geographic area of the fishery. As wreckfish flooded the market, dock prices given fishers fell to record lows. By 1991, ninety vessels had obtained permits for a fishery that had begun just four years before.

Tracking compliance with quotas in the derby fishery was very difficult in several ways (Gauvin 1994). Because wreckfish were off-loaded in many ports from North Carolina to Florida, often were trucked across state lines, and were processed, sold, and resold, hundreds of establishments were involved in the trade. Incentives for compliance were low, and many businesses were not included in the data collection system. The consequences of this underreporting included possible quota-busting and deterioration in the quality of information used in assessing the status of the stock.

The ITQ Program

Unlike previous regimes at the National Marine Fisheries Service, the regime at the end of President George Bush's administration was sympathetic, even encouraging, toward SAFMC consideration of access controls and market-based approaches to fishery management. The South Atlantic Fishery Management Council began discussing an ITQ program for the wreckfish fishery in 1991 (Gauvin, Ward, and Burgess 1994). Along with concerns about economic waste in the fishery, the discussion focused on ITQs as a means of decreasing conflicts among fishers and creating incentives for compliance with fishery regulations. The SAFMC also was convinced that an ITQ program would promote a conservation ethic and regulatory compliance by vesting fishermen in the fishery.

The wreckfish fishery also had several characteristics that made it attractive for ITQ management (Gauvin 1994). For instance, by comparison with other fisheries in the region that had as many as 1,500 permit holders, the wreckfish fishery had relatively few participants. Also, the fishery targeted a single species that was not also sought by recreational fishers.

The SAFMC adopted an ITQ program for the wreckfish fishery for the 1992–1993 fishing season (Gauvin, Ward, and Burgess 1994). As in other fisheries considering ITQ programs, a key controversy concerned initial allocation. Fishers who had participated in the fishery from the beginning argued for basing allocation entirely on catch history. Fishers who had recently entered the fishery argued that the shares should be allocated equally. In the end, the SAFMC decided to use a formula weighted partially on catch history. To be eligible, a vessel owner had to present fish house receipts and affidavits attesting to landings of at least 5,000 pounds in either 1989 or 1990.[49] Of a total of 100 shares in the fishery, 50 were divided equally among eligible participants and the rest were divided among participants on the basis of documented catch in 1987–1990.

No single business could receive more than 10 of the 100 shares initially (Gauvin, Ward, and Burgess 1994). This limitation reflected several concerns. For one thing, it was feared that a business receiving a large initial share would enjoy an advantage in purchasing shares from other participants. Also, the limitation was meant to address uncertainty over the fairness of the formula used in the initial allocation.

The SAFMC debated whether to place a cap on the number of shares any one vessel owner could hold (Gauvin, Ward, and Burgess 1994). Some fishers were concerned that without a cap, a small number of businesses could capture the majority of shares and control prices. Others argued that the shares provided fishers with protection from unrestrained entry into a valuable fishery. Since wreckfish were only part of the overall supply of grouperlike fish in the market, control of the supply of wreckfish would not affect prices. In the end, the SAFMC decided to leave questions of consolidation to antitrust law.

As in other fisheries under ITQ management, concern arose in the wreckfish fishery that fishers receiving shares were benefiting from a windfall (Gauvin, Ward, and Burgess 1994). Many members of the SAFMC believed that shareholders should pay for administration of the program if they were going to profit as expected. However, the Magnuson-Stevens Fishery Conservation and Management Act restricted fees to those administrative costs associated with issuing permits, which amount to about $53 per year. The final ITQ management plan stated that if Congress were later to provide the authority, the SAFMC would impose fees to cover the costs of tracking and monitoring the ITQ shares.

Although the ITQ program authorized the later purchase, lease, or rental by anyone of shares or portions of shares, the SAFMC sought to avoid giving shares legal status as property. Among other things, the plan stated that the shares were of indefinite duration and that the SAFMC could rescind the ITQ program if it found that the program was not meeting its objectives (Gauvin, Ward, and Burgess 1994).

Each year, shareholders are given catch coupons (issued in denominations of 500 and 1,000 pounds) equal to their share of the overall quota (Gauvin, Ward, and Burgess 1994). On selling wreckfish to a dealer, a fisher gives the dealer canceled coupons equal to the weight of the landings. Shareholders may transfer coupons among themselves.

Early Performance of the ITQ Program

In February 1992, 49 vessel owners were given shares in the wreckfish fishery (Gauvin, Ward, and Burgess 1994). By 1993, 23 shareholders had sold their shares and 5 new shareholders had entered the fishery. The number of vessels permitted for the fishery also fell, from 91 in the 1991–1992 season, which ran from April through March, to 40 for the 1992–1993 season. Only 20 vessels actually fished for wreckfish.

These changes partly reflected the fact that for most vessel owners, wreckfish would never be a primary fishery. Instead, many fishers who had entered the wreckfish fishery moved among several fisheries, including snapper, grouper, and shrimp (SAFMC 1993). Depending on how well these other fisheries were faring, wreckfish ITQ shareholders would transfer their shares or coupons and focus their fishing elsewhere. This led to a slight concentration of shares quite quickly, but it was far from a monopolistic hold on the fishery (Gauvin, Ward, and Burgess 1994).

Other changes in the fishery appeared early (Gauvin, Ward, and Burgess 1994). For instance, in the 1991 derby fishery, fishers focused their effort early in the year, when wreckfish are more dispersed and catch rates are low. In 1992, the number of trips in April fell by more than half, to seventeen. In 1993, April trips fell again, to just ten. Landings also fell, to 1.3 million pounds in 1992 and 1.1 million pounds in 1993 (SAFMC 1997) (see table 6.2).

Among other things, the more even distribution of lower landings over the year averted the gluts that marked the derby fishery. Dock prices, which had ranged between $0.90 and $1.55 per pound, stabilized early at a higher level, about $1.85 per pound (Gauvin, Ward, and Burgess 1994). Linking of this higher, more stable price to the ITQ program, however, would require empirical evidence that did not exist. Still, such a price increase in an open access fishery would have attracted more vessels and fishing effort.

Whether or not the ITQ program itself increased fishers' commitment to conservation, arguments for setting TACs at higher levels vanished (Gauvin, Ward, and Burgess 1994). Establishment of a TAC for a species so poorly known as wreckfish has its risks and uncertainties. In the past, some fishers, pointing to lack of evidence for an impending decline in wreckfish abundance, had argued for setting the TAC as high as 6 million pounds (Gauvin 1994). When asked to comment on a proposed TAC of 2 million pounds after the ITQ program began, most fishers said that the wreckfish stock could sustain a higher catch.

Table 6.2. Wreckfish Season Comparisons, 1991–1992 through 1996–1997

Season	*Vessels*	*Trips*	*Days*	*Pounds*	*Fish*	*Size*	*Catch per Trip*	*Days per Trip*	*Catch per Hour*	*Catch per Day*
1991–1992	38	308	2,164	1,962,088	57,704	33.35	6,254	7.03	196.92	983.78
1992–1993	20	222	1,516	1,270,557	38,205	33.20	5,723	6.83	183.56	931.57
1993–1994	19	210	1,531	1,144,726	33,803	33.97	5,451	7.29	152.32	758.26
1994–1995	17	201	1,602	1,203,265	35,770	34.11	5,986	7.97	193.04	813.75
1995–1996	13	140	946	644,997	19,256	33.67	4,607	6.76	218.40	750.52
1996–1997	9	95	762	396,868	11,657	34.21	4,178	7.84[a]	164.24	547.84[a]

[a] Excludes a 25-day trip that included a major breakdown. Inclusion of the trip would have increased mean days per trip to 8.02 and reduced mean catch per day to 542.74.
Source: Hardy 1997.

But none pushed for a higher quota. Instead, they maintained that they could make a living with the proposed TAC, since they would get a higher price for their catch and spend less on catching it. They were no longer willing to accept the risk of overfishing.

Another piece of anecdotal evidence for a conservation ethic in the fishery involves changed fishing practices. Before the ITQ program, fishers regularly deployed bottom longlines because they were so efficient in catching wreckfish. Knowing that the cables could cause great damage to the coral heads and ledges typical of wreckfish habitat, fishers argued that they would not use such gear if other fishers did not use it. Fishers regularly ignored a regulatory prohibition on the use of bottom longlines. However, once the ITQ program was under way, fishers began reporting suspected use of bottom longlines to enforcement agents. Compliance with the prohibition was high, partly, no doubt, because violation of the prohibition would lead to forfeiture of quota shares.

Enforcement agents found that the percentage of fishers in compliance with rules in the fishery rose with the ITQ programs (Gauvin 1994). Although fishers might seek to benefit by catching more than they reported, there were few incidents of quota-busting or high-grading, practices that undermine the conservation value of a TAC (Gauvin, Ward, and Burgess 1994).

Although a conservation ethic may have played a role in bringing about compliance with quotas, the design of the coupon system itself also contributed. Enforcement officials designed the coupons so that they could not be filled out quickly by a fisher wishing to avoid detection of a violation by an approaching enforcement agent. High monetary penalties and possible forfeiture of shares and permits discourage violation of quotas by fishers and misreporting by dealers (Gauvin 1994). The relatively low price of additional shares or coupons probably also militates against violation of quotas.

High-grading has been absent from the fishery as well (Gauvin, Ward, and Burgess 1994). The ITQ program itself probably played little role in this. Rather, high-grading has been low largely because the market does not pay a premium for wreckfish of certain sizes. Indeed, wreckfish landed by fishers are almost all the same size (Gauvin 1994).

Besides helping to ensure compliance with the overall TAC, the ITQ tracking system improved the reliability of catch information used in managing the fishery (Gauvin 1994).

Recent Performance of the ITQ Program

The trading and ownership of shares since the beginning of the ITQ program in 1992 shows some consolidation in the fishery. Between 1992 and 1997, a little more than 58 shares were traded in twenty-five transactions, at an average value per share of $9,787.14 (SAFMC 1997). The number of shareholders in the fishery fell from 49 at initial allocation to 25 in the 1996–1997 season. The average number of shares held by shareholders rose from about 2 at initial allocation to 4. The percentage of shares held by the largest 5 shareholders rose from about 24 percent to more than 56 percent. Clearly, there has been some consolidation of shares in the fishery.

Other indicators show a decline in active participation in the fishery. The percentage of shareholders who did not use coupons rose from 23 percent in the 1993–1994 season to 60 percent in the 1995–1996 season. In the 1995–1996 season, no shares were traded, no coupons were purchased, and no shareholder used all of the allocated coupons (see table 6.3). The number of vessels that participated in the fishery fell from 19 in 1993 to 9 in 1996 (table 6.2). In 1996, the number of trips was less than half what it was in 1993, as was the number of days spent fishing.

Between 1993 and 1996, the number of fish landed fell from

Table 6.3. Coupon Utilization by Shareholders

Season	*Number Shareholders*	*Used 0–50% of Coupns*	*Used 51–99% of Coupons*	*Used >100% of Coupons*	*Sold All or Part of Coupons*	*Did Not Use Coupons*
1992–1993	38	21	11	16	26	26
1993–1994	26	23	31	8	15	23
1994–1995	25	15	15	15	19	35
1995–1996	25	20	20	0	0	60

Source: SAFMC 1997.

33,803 to 11,657, and the weight of the catch declined from 1.1 million pounds to 396,868 pounds, all well below the TAC of 2.0 million pounds (table 6.2). The average weight of individual wreckfish has remained at about thirty-four pounds. Between 1994 and 1997, the price per pound for wreckfish increased from $1.83 to $2.10 per pound, reflecting declining landings (SAFMC 1997).

The reasons for several of the decreases are unclear. The decline in number of trips, number of landings, and coupon use may be due to weather or to heavier participation in other fisheries, such as shark, swordfish, snowy grouper, and tilefish (Gauvin 1994; SAFMC 1997). The decline in catch per day from about 758 pounds in 1993 to about 548 pounds in 1996 may be temporary, but continued decline in catch per day would indicate possible overfishing (SAFMC 1997). Biologists have also expressed concern about uncertainty regarding the biology of wreckfish and the effects of fishing in other parts of the North Atlantic.

Whatever the mix of reasons for the decline in fishing effort, wreckfish are experiencing a respite enjoyed by few other populations of fish.

Halibut and Sablefish in Alaska

The following case illustrates what can happen in a fishery managed by catch limits only. In chapter 5, we explained what happens when a TAC is set below the initial capacity of the fleet. Capacity catches up, the stock is pressured, and the TAC is reduced. Individual skippers face the possibility of the season ending before they reach their desired catch, that is, before they cover all their fixed costs. Some may leave the fishery, but others will remain and may increase their investment in order to raise their portion of the TAC in what becomes a shorter and shorter season. A derby-style fishery ensues, pursued by a larger number of more intensely capitalized vessels fishing faster and faster. This example also illustrates what happens when an individual quota system is introduced into a derby fishery.

The individual fishing quota (IFQ) program established in 1995 for the sablefish and halibut longline fisheries off Alaska is by far the most ambitious and controversial such program in the United States. Between 1985 and 1990, the annual number of vessel owners in the halibut fishery alone ranged between 2,479 and 3,883, nearly 90 percent of whom were Alaska residents (Pautzke and Oliver 1997). In the sablefish fishery, 1,094 vessel owners eventually qualified to receive initial allocations of quota shares.

Pacific halibut *(Hippoglossus stenolepis)* are found from the Sea of Japan to the Bering Sea and southward as far as Baja California, although they are rare south of San Francisco (Love 1996). There may be two separate populations, one found from the Bering Sea westward to Japan and the other from the Gulf of Alaska southward. Females mature at eight to sixteen years of age, males at five to thirteen years. Pacific halibut can live as long as forty-two years and reach a length of more than eight feet. Although some Pacific halibut are taken by recreational anglers, most are caught commercially on longlines in waters 100–750 feet deep.

Landings of Pacific halibut have fluctuated widely (see figure 6.7). Landings by U.S. and Canadian fishermen rose from 14,158 metric tons in 1950 to 48,382 metric tons in 1962 and then fell to a new low of 13,075 metric tons in 1974. By 1988, landings had risen again to a peak of 44,741 metric tons before falling to 25,965 metric tons in 1995. In that year, the U.S. fleet landed 20,551 metric tons of Pacific halibut.

Since 1982, fisheries for halibut have been jointly managed by the International Pacific Halibut Commission (IPHC), established in 1953,[50] and two regional fishery management councils, the North Pacific Fishery Management Council (NPFMC) and the Pacific Fishery Management Council (PFMC), established in 1976 by the Mag-

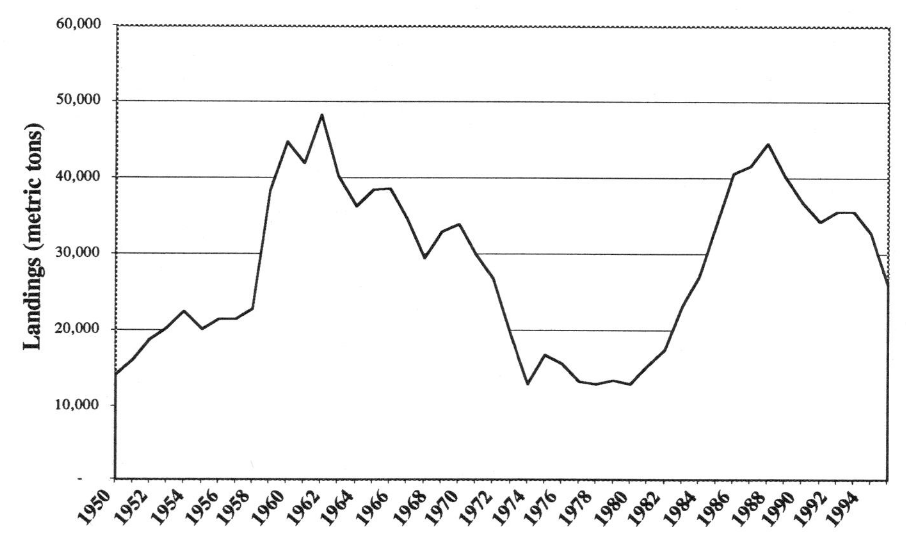

Figure 6.7. Landings of Pacific Halibut from the Northeastern Pacific (*source:* FISHSTAT 1997)

nuson-Stevens Fishery Conservation and Management Act (Pautzke and Oliver 1997). The IPHC is responsible for biological management of Pacific halibut, including the setting of total allowable catches (TACs), and the regional fishery management councils are responsible for allocating TACs.

The NPFMC began developing a fishery management plan for halibut as early as 1978 (Pautzke and Oliver 1997). The draft plan included a controversial proposal to set a cutoff date for accruing credit under any future limited entry program. Longtime fishermen, many of whom were based in Seattle, Washington, favored the program, and fishermen with less history in the fishery opposed it.

As more fishers entered the fishery, partly in anticipation of limited entry, fishing seasons grew shorter (Pautzke and Oliver 1997). By 1980, the season for fishing in one zone off Kodiak, Alaska, had fallen from 150 days to fewer than 100. By the late 1980s, fishing was confined to two twenty-four-hour seasons each year. In a scene reminiscent of a horse race, fishing vessels would line up, awaiting the opening of the season, and then rush out, regardless of weather, to catch as much as they could as quickly as they could. Halibut flooded the processing plants, so most of the catch had to be frozen for later distribution.

In 1982, the NPFMC proposed a moratorium on issuance of new halibut licenses while it studied proposals for limiting entry (Pautzke and Oliver 1997). The National Marine Fisheries Service and the Office of Management and Budget (OMB) rejected the proposal. Among other objections, the OMB argued that a moratorium would not address overcapitalization in the fishery and would create new inefficiencies by preventing the entry of fishers who might catch and market halibut profitably. The NPFMC dropped the proposal and did not take up the issue of limited entry for several years.

Like Pacific halibut, sablefish *(Anoplopoma fimbria)* range from Japan to the Bering Sea and southward to Baja California (Love 1996). Sablefish school in deep water over sand or mud. They may reach a length of forty inches and live as long as fifty-five years, although they mature as early as five years of age. Sablefish are caught in the Gulf of Alaska in trawls, fish traps, and longlines. Much of the catch is exported to Japan, where sablefish are graded by their oil content for use as sashimi.

Like landings of Pacific halibut, landings of sablefish have fluctuated widely (see figure 6.8). Landings rose from 1,401 metric tons in 1958 to 59,000 metric tons in 1972 and then fell to 17,662 metric tons in 1980. Landings of sablefish reached a new peak in 1988, at 51,678 metric tons. By 1995, landings had fallen to 31,108 metric tons. All but 3,838 metric tons of this was landed at U.S. ports; the balance was landed at Canadian ports.

Until the 1980s, nearly all sablefish were caught by foreign trawlers and long-liners (Pautzke and Oliver 1997). By 1984, a small fleet of U.S. long-liners was landing 2,000 to 3,000 metric tons of sablefish, mostly from the eastern part of the Gulf of Alaska. By 1987, when foreign fishing for sablefish ceased, U.S. long-liners were landing 37,600 metric tons of sablefish, two-thirds of it from the Gulf of Alaska.

The fleet grew in number, size, and sophistication of vessels (Pautzke and Oliver 1997). From 1981 to 1988, the number of vessels longer than fifty feet grew tenfold and the number of smaller vessels grew by a factor of fourteen. By 1992, roughly 1,000 vessels were fish-

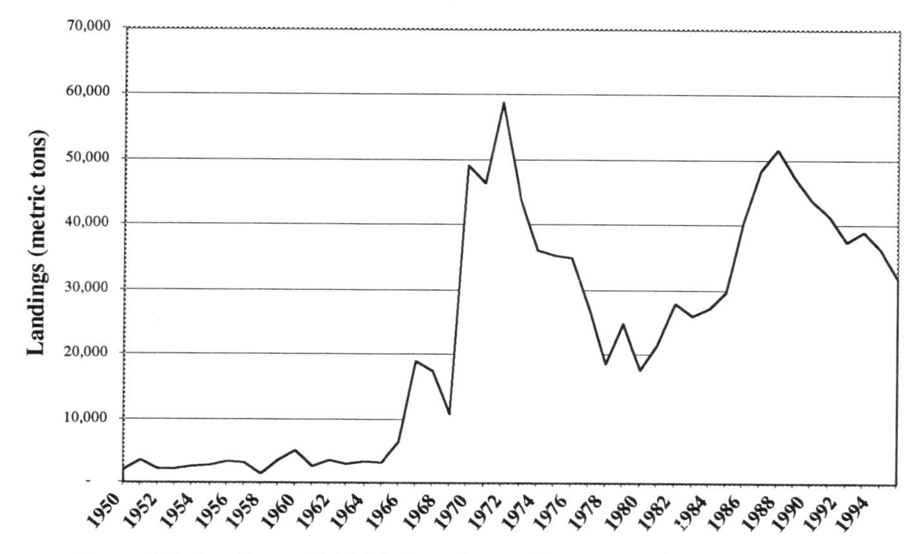

Figure 6.8. Landings of Sablefish from the Northeastern Pacific (*source:* FISHSTAT 1997)

ing for sablefish. As the fleet grew, the fishery expanded westward and seasons shortened. Between 1984 and 1990, the season in the eastern part of the Gulf of Alaska fell from 180 days to 20 days. The fishing season in the central gulf fell from 254 days to 60 days.

In 1987, the NPFMC began considering the use of alternative management measures in the sablefish fishery, including license limitation and use of individual transferable quotas (Pautzke and Oliver 1997). It soon added the halibut fishery to its consideration as well. (By 1990, fishing for halibut was restricted to three or four one- or two-day seasons.)

In December 1991, the NPFMC selected individual fishing quotas (IFQs) as the preferred method for managing the longline fisheries for halibut and sablefish (Pautzke and Oliver 1997). As in other ITQ-like programs, the halibut and sablefish IFQ program aimed at enabling fishers to adjust the time and intensity of fishing and to focus on efficiency and product quality rather than speed.

The NPFMC awarded initial quota shares to vessel owners and leaseholders, partly because they had already placed their effort and capital at risk in the fishery (Pautzke and Oliver 1997).[51] Crew members who could demonstrate a history in the fishery were given priority in future transfers of shares.

Quota shares are valid indefinitely; however, if the NPFMC rescinds the IFQ program, shareholders will not receive compensation. The NPFMC separately set aside portions of the quotas for halibut and sablefish for Native Americans living in villages of western Alaska.

To qualify to receive a quota share, a vessel owner had to have landed halibut or sablefish at least once during the 1988–1990 season (Pautzke and Oliver 1997). The number of shares allocated was based on catch history: the best five of the six years during 1985–1990 for sablefish and the best five of the seven years during 1984–1990 for halibut. Generally, fishers who had fished during only one year received shares equal to about 20 percent of their catch. If they had fished during two years, they received roughly 40 percent of their annual catch. Fishers who had fished during five years received roughly their average annual catch. The number of pounds of fish that a share equals fluctuates with the annual total allowable catch.

Concerns about concentration of shares in the hands of larger, wealthier operators arose here as in other fisheries proposed for ITQ management (Pautzke and Oliver 1997). To promote the persistence of owner-operated fleets in these fisheries, the NPFMC adopted several measures. First, eligible vessels were divided into two categories: freezer long-liners that remain at sea for long periods and smaller catcher boats that deliver their iced fish to shoreside processors. Categories for catcher boats were further determined by vessel length, with two size categories in the sablefish fishery and three in the halibut fleet.[52] The NPFMC also limited the percentage of total shares that any one owner could hold to 0.5 percent or 1.5 percent of the total, depending on the fishery and the area.

Other measures restricted the lease and sale of shares, particularly in the first three years of the program (Pautzke and Oliver 1997). Crew members were permitted to purchase catcher boat shares, but corporations were allowed to do so only if they received an initial allocation. Owners of trawlers were not permitted to purchase shares. Only vessels holding quota shares were allowed to land halibut or sablefish.

The NPFMC established a more extensive monitoring and

enforcement system than is found in the ITQ programs for wreckfish and for surf clams and ocean quahogs on the Atlantic coast. All sales, transfers, and leases must be approved by the Department of Commerce and are monitored by the National Marine Fisheries Service (Pautzke and Oliver 1997). Before landing their catch, fishers must notify the NMFS; the catch is then logged against the individual quota. Up to a certain point, catch that exceeds the IFQ amount is deducted from the following year's individual quota.[53] Catch that exceeds this amount triggers other penalties.

The NPFMC adopted the IFQ program only after extensive public hearings and analysis of the likely effects of the program. Some expected benefits could not be quantified (Pautzke and Oliver 1997). For instance, safety was likely to increase because fishers would not be forced to catch their entire quota in a short period of time. Since landings would be spread out over the season, opportunities for employment in processing would increase (Pautzke and Oliver 1997).

The NPFMC's analysis predicted other benefits as well. For instance, the value of the catch, and hence the prices paid to fishers, would probably increase because market gluts would be avoided and more of the catch could be marketed fresh rather than frozen (Pautzke and Oliver 1997). Halibut caught in the Canadian fishery, which had operated under an ITQ program for several years, regularly sold in supermarkets for $10.99 per pound, compared with $4.99 for frozen Alaska halibut. Analysts expected that the higher quality of Alaska halibut resulting from an ITQ program would generate between $4.8 million and $38.5 million, including $3.1 million in reduced costs for storing frozen halibut.

Other expected economic benefits included a savings of $1.2 million to $2.0 million from reduced mortality of halibut and sablefish caught in lost fishing gear (Pautzke and Oliver 1997). In 1990, fishers lost an estimated 1,860 skates (a standard unit of line of length with spaced hooks) of longline worth $3.0 million, killing nearly 2 million pounds of halibut. Savings of $12.4 million to $13.6 million were expected to result from more efficient fishing strategies and redistribution of fishing effort to lower-cost fishing operations.

In all, analysts estimated that total annual benefits would range between $30.1 million and $67.6 million (Pautzke and Oliver 1997).

Another $11.0 million to $13.9 million in benefits would have been expected had the IFQ program not restricted the transfer of quota shares among vessel categories.

Fishers clearly see long-term value in holding quota shares. In 1995, shares of halibut quotas sold for prices ranging from $5.47 per pound to $8.17 per pound (Parker and Macinko 1996). Average prices for sablefish quota shares ranged between $4.21 per pound and $6.28 per pound. Lease rates ranged between $0.80 and $1.04 per pound for halibut and between $0.53 and $1.05 for sablefish.

Although the question of whether quota shares are private property remains unsettled, private banks and government agencies have come to treat quota shares as having financial value that may allow them to serve as collateral for loans, for instance. By the end of 1997, banks had placed liens on 388 quota shares; private lenders, on 116 shares; the Internal Revenue Service, on 79 shares; and Alaska's child support program, on 26 shares.[54]

Transfer of shares has been vigorous. In 1997, the NMFS processed 1,911 requests for transfer of shares, nearly the same number as in the previous two years.[55] Nearly half of these transfer requests were permanent.

The transfer of quota shares has led to some consolidation of shares. Although the number of shareholders with halibut shares less than 10,000 pounds declined by 31 percent, those with shares greater than 10,000 pounds increased by 5 percent[56] (see figure 6.9). In the sablefish fishery, the number of shareholders with shares less than 10,000 pounds fell by 22 percent and the number of those with shares greater than 10,000 pounds also fell, by 2 percent (see figure 6.10).

Crew members who can demonstrate that they have served as crew in any U.S. fishery for 150 days may lease or purchase quota shares.[57] Since the beginning of the ITQ program, 1,573 crew members have qualified, and nearly half of these have obtained shares, accounting for 11.3 percent of the halibut quota and 4.5 percent of the sablefish quota.

The number of vessels reporting landings of halibut and sablefish have declined dramatically. In areas 2C and 3A, which account for about three-quarters of the landings, the number of vessels reporting landings fell by 44 percent between 1992 and 1997[58] (see figure 6.11).

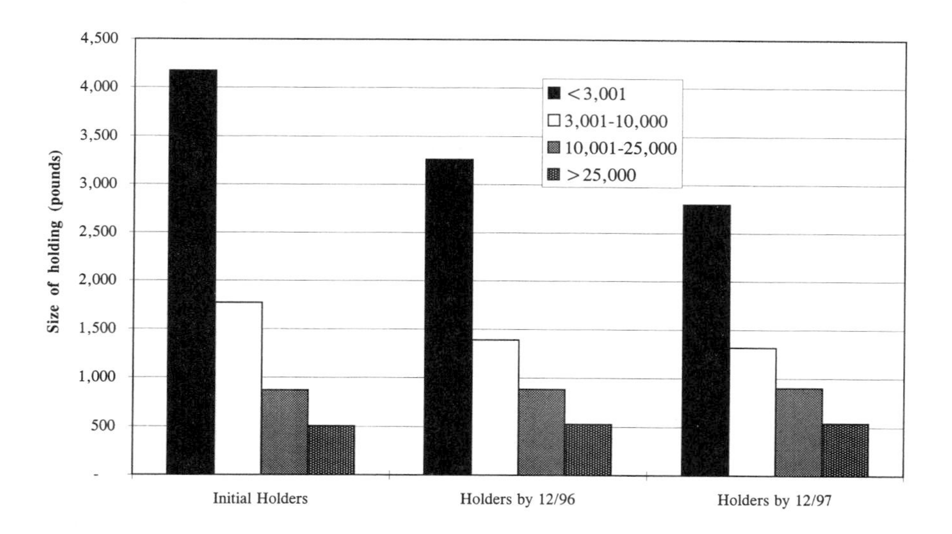

Figure 6.9. Number of Persons Holding Halibut Quota Shares, by Size of Holding in Pounds (*source:* P. Smith, letter to Richard Lauber, chairman, North Pacific Fishery Management Council, December 1997) [Internet: <http://www.fakr.noaa.gov/ifq/dec1997.htm>]

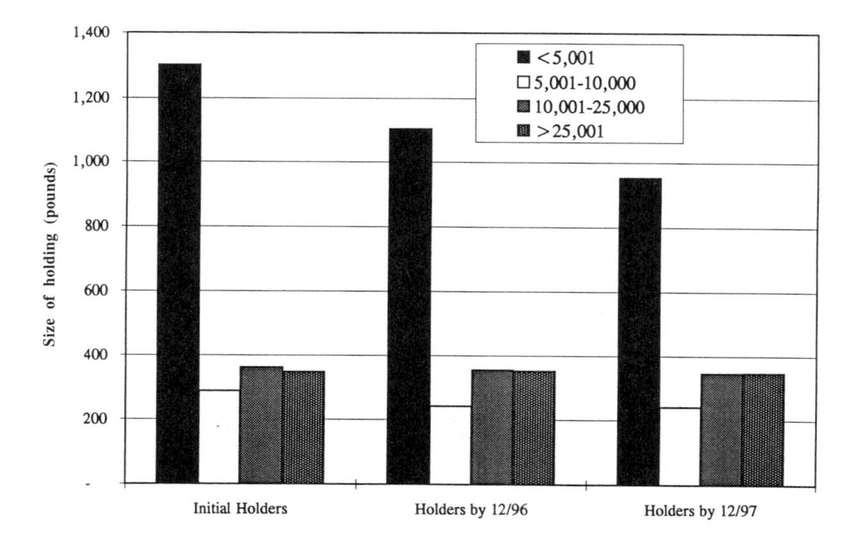

Figure 6.10. Number of Persons Owning Sablefish Quota Shares, by Size of Holding in Pounds (*source:* P. Smith, letter to Richard Lauber, chairman, North Pacific Fishery Management Council, December 1997) [Internet: <http://www.fakr.noaa.gov/ifq/dec1997.htm>]

The number of vessels reporting landings declined in all other areas except 4D. The International Pacific Halibut Commission manages halibut fishing by adopting measures such as open and closed seasons and quota limits in each of ten regulatory areas from California to the Bering Sea. Areas 2C and 3A cover the water off southeastern Alaska and the central Gulf of Alaska. In the sablefish fishery, the number of vessels reporting landings in the three major fishing areas declined by 26–54 percent between 1992 and 1997 (see figure 6.12).

The IFQ program has also reduced bycatch and discards of halibut, in the sablefish fishery in particular (Pautzke and Oliver 1997). In the past, sablefish fishers were required by regulation to discard any halibut captured incidentally on their longlines. Bycatch of halibut was common enough that the NPFMC set a 750-ton cap on bycatch and discards of halibut in the sablefish fishery. Sablefish fishers who also held

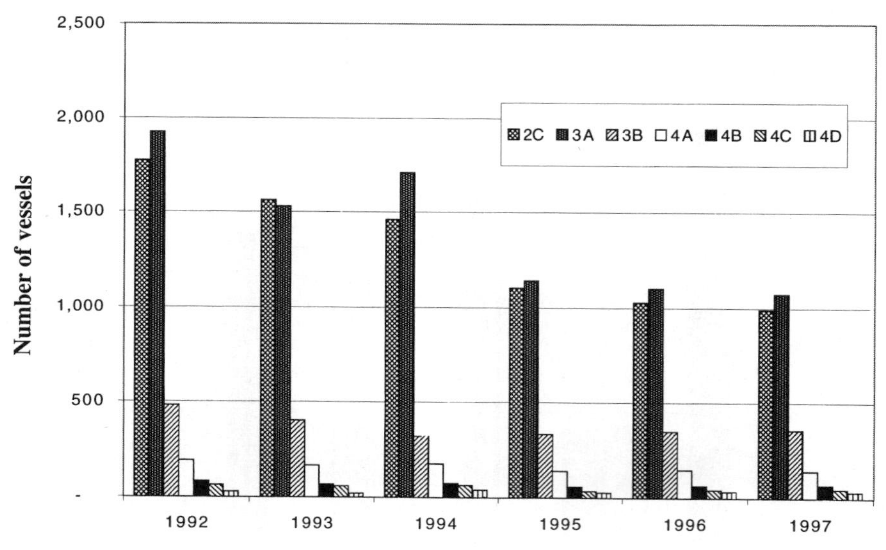

Figure 6.11. Number of Vessels Reporting Halibut Landings, by Area (*source:* P. Smith, letter to Richard Lauber, chairman, North Pacific Fishery Management Council, December 1997) [Internet: <http://www.fakr.noaa.gov/ifq/dec1997.htm>]
note: 2C = Southeast Alaska; 3A = Northern Gulf of Alaska; 3B = Western Gulf of Alaska; 4A, B, C, D = Bering Sea/Aleutian Subareas

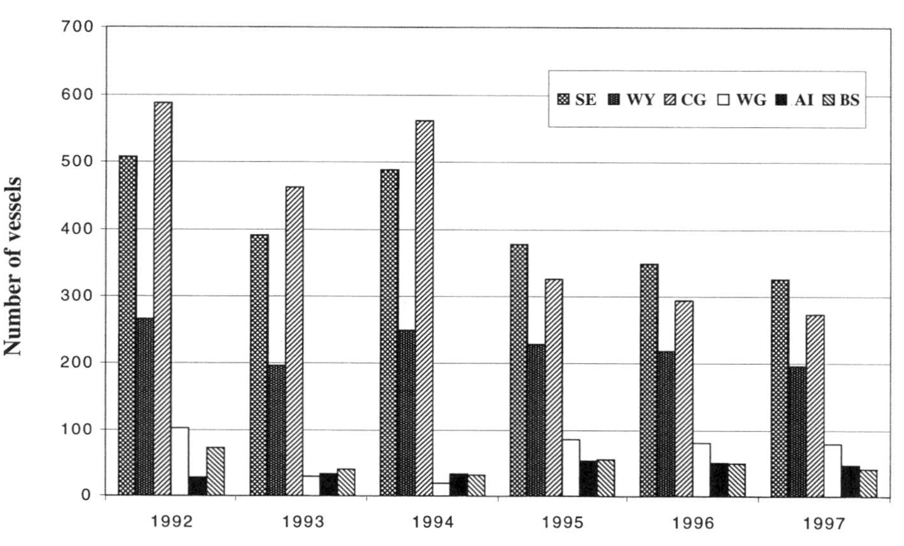

Figure 6.12. Number of Vessels Reporting Sablefish Landings, by Area (*source:* P. Smith, letter to Richard Lauber, chairman, North Pacific Fishery Management Council, December 1997) [Internet: <http://www.fakr.noaa.gov/ifq/dec1997.htm>] *note:* SE = Southeast, WY = West Yakutat, CG = Central Gulf of Alaska, AI = Aleutian Islands, BS = Bering Sea

IFQ shares for halibut have been able to retain their halibut bycatch. As a result, the cap has been reduced to 150 tons annually in the Gulf of Alaska.

The controversy that attended the review and implementation of the ITQ program in these fisheries led to establishment of an inter-agency team to evaluate the effects of the program (Parker and Macinko 1996). The team evaluated the program's distributional effects and effects on conservation and conducted interviews with shareholders and buyers. Among other things, this review of the program's first year of operation found the following:

- Most people receiving halibut quota shares were relatively small holders. Corporations held a much higher proportion of sablefish quota shares.

- The International Pacific Halibut Commission estimated that lost or abandoned gear killed far fewer halibut than in the past.
- No significant underreporting of catches or high-grading was found.
- Most processors said they paid higher ex-vessel prices for halibut and sablefish and received higher wholesale prices.
- Many halibut shareholders fish together with other shareholders.
- The most commonly mentioned positive effects of the program included the ability to choose when to fish, increased safety, better markets and prices, uncrowded fishing grounds, and a more relaxed fishery.
- The most commonly mentioned negative effects included small or uneconomic quota share allocations, administrative problems, and unfair quota allocations.
- The number of U.S. Coast Guard search and rescue cases fell by nearly half, from an average of twenty-seven each year before implementation of the IFQ program to fifteen in 1995, and the number fell further in 1996 (Welch 1997).

Because widespread failure to report landings fully would reduce the value of shares, effective enforcement of the ITQ program is critical. The NMFS estimated that the minimum enforcement effort would require an increase in the number of enforcement staff members from twenty-eight to sixty-two in the Alaska region (Pautzke and Oliver 1997). But NMFS proposals to fund such an increase have been rejected by Congress. As a result, only a small fraction of landings have been monitored in the halibut and sablefish fishery since the ITQ program became effective in November 1995. Enforcement investigations indicated that one in five participants in the fishery were in violation of regulations (Parker and Macinko 1996). The effect of this level of enforcement is unknown.

Less formal reports have identified other changes in the fishery. Some processors and distributors have lost revenues and jobs as more fish was shipped fresh by air from Alaska to markets in the lower forty-eight states rather than being frozen and shipped by truck (Welch 1997). Fishers have gained leverage in bargaining for prices as gluts have disappeared.

Canada's Pacific Halibut Fishery

The halibut fishery of the North Pacific, one of the oldest commercial fisheries in the region, is managed under a treaty between the United States and Canada through the International Pacific Halibut Commission (IPHC).[59] More than 4,500 U.S. and 435 Canadian vessels are licensed to participate in the fishery, which has been operating since the late 1800s. The IPHC determines biomass and stock strength, sets an annual total allowable catch (TAC) for Canada and the United States, and establishes fishing seasons. The two participating countries manage and enforce the regulations their governments establish. In 1994, the United States and Canada landed 27,100 tons and 6,000 tons of halibut, respectively, worth $114.4 million (IPHC 1997).

The fishery occurred throughout the North Pacific until 1977, when both Canada and the United States declared 200-nautical-mile exclusive economic zones and fishers were restricted to their own national waters. In 1979, as repatriated Canadian vessels targeted the limited stock in their own waters, Canada's Department of Fisheries and Oceans (DFO) moved to limit entry into the fishery. Although the limited entry scheme was intended to stop expansion of catch capacity, protect the halibut stocks, and enhance economic benefits to the fishermen, it soon fell short of these goals. The licenses took on significant value and were traded among fishermen, who then, to justify their investment, fished harder and more efficiently.

In order to compete more effectively, vessel owners turned to bigger boats, larger crews, round-the-clock fishing, more gear, sounders, sonar, loran, circle hooks, snap-on gear, automatic baiters, and hook strippers. The race to fish took on the nature of a derby, with the season shrinking from sixty-one days in 1982 to only six days in 1990. The compressed season and long working hours endangered crew and vessels, intensive fishing in all weather caused lost and damaged gear, and the breakneck pace meant less time spent in handling the catch properly. The shorter fishing periods also meant that Canadian fishermen were able to target the lucrative market for fresh fish only once or twice a year, leaving the majority of the landed catch for the lower-value frozen fish market. As the openings became shorter, financial losses associated with vessel breakdown, sickness, injury, and poor weather represented substantial reductions in earnings.

During these brief season openings, millions of pounds would be landed in less than a week, resulting in fish sitting on the dock for days before being shipped to market or placed in cold storage. Not only was the directed catch poorly handled, but also bycatch species were discarded and rarely documented in catch records. The DFO was concerned about the amount of fish discarded during the derby openings, such as rockfish, lingcod, dogfish, and sablefish. Either prohibitions on retention or low economic value meant that most, if not all, of this bycatch was thrown back, most of it dead.

Although the season lasted only six days, licensed halibut vessels often set gear early and hauled after the closure. A budget-strapped DFO was unable to monitor the fishery or check landings, relying instead on the honesty of fishermen and processors.

The IPHC was finding it difficult to manage the fishery. In the decade prior to 1991, the catch had consistently exceeded the TAC by an average of 3.4 percent. Between 1987 and 1990, the average annual overage was 5 percent, and in 1990 alone the catch exceeded the TAC by 10 percent.

In 1988, a local fishing vessel organization established a small subcommittee to study alternative management approaches for the halibut fishery. The group focused on individual quotas (IQs) and approached the DFO for assistance in developing a concept for the halibut fishery. Another organization joined the effort, and at meetings in the spring of 1989 the two organizations discussed how an IQ program might work in the halibut fishery. A discussion paper was distributed throughout the industry, generating interest and controversy and prompting the DFO to set up industry meetings.

The fleet turned out for the meetings, and in response to a survey, the fishers resoundingly informed the DFO of their interest in proceeding with an IQ approach. They made clear, however, the need for a Halibut Advisory Board (HAB). The industry did not have a formal advisory process and representation was a mixed affair, with mostly salmon organizations representing gear types, regions, and vessels. In addition, there were independent fishers, Native Americans, and processor-owned boats. The challenge was to create an advisory board of license holders, processors, Native Americans, and union representatives that would not disenfranchise anyone, would be comprehensive in its representation, and would be democratic and accountable.

The industry recommended, and the DFO carried out, a scheme of representation whereby every representative of license holders to the HAB had to obtain twenty signatures from registered halibut license holders, each of whom could align with only one representative. Each license-holder representative had a list of supporters to whom the representative was accountable. Processing, Native American, and union representatives also were asked to participate.

The first meeting took place in January 1990 and lasted four days as participants struggled to find an acceptable initial allocation formula. An impasse threatened to stop the entire process when negotiations resulted in two approaches. But brinksmanship by the DFO, which threatened to terminate the entire IQ process if no compromise could be reached, resulted in a deal. Industry representatives emerged with an allocation formula they had negotiated democratically with almost unanimous support from license-holder representatives.

During the following months, the HAB met to discuss other issues, such as transferability, cost recovery, enforcement, reporting requirements, landing sites, appeals, and season length. By June, the HAB had a complete halibut IQ proposal to send to each license holder and had outlined a two-year trial program. Seventy percent of license holders voted to accept the proposal, although processing companies and the union representing crew members opposed it.

The minister of fisheries and oceans approved the plan in November 1990 and announced that the two-year trial period of IQ management would begin in 1991. The HAB met many times more before the start of the 1991 fishing season in order to resolve operational problems and minimize disruptions during the transition.

Key Features of the Program

Participants in the advisory process set guidelines for the design of their IQ program: keep it simple so fishers could understand the rules, find it easy to cooperate, and determine whether or not IQs were a better management method; keep it flexible so the program could adapt to unanticipated changes and needs; keep it open so as not to disenfranchise historical participants in the halibut fishery; and make monitoring and enforcement comprehensive and effective.

The management plan had the following features:

1. A two-year trial period was established encompassing the 1991 and 1992 halibut seasons, to be open from March through October.
2. An allocation formula awarded each vessel a percentage of the annual TAC. Seventy percent of the allocation was based on the vessel's performance between 1986 and 1989, and 30 percent was determined by the vessel's length. In 1991, IQs ranged from 4,000 pounds to 70,000 pounds. License holders who disagreed with the share established for their vessel had the right to appeal to an independent Halibut Quota Review Board.
3. Transferability was allowed, but not quota-stacking during the trial period. The IQ was attached to the halibut vessel license, and each vessel license had a specific size restriction. A halibut license and quota could be transferred to another vessel of similar size as long as the vessel was not already associated with a halibut license.
4. A monitoring and enforcement plan was established that included unlicensed trawlers, trollers, and groundfish hook-and-line vessels as well as licensed halibut vessels.

Before fishing for halibut, fishermen were to advise the HAB of their intention to do so, their intended fishing location, and their expected landing date and port. Prior to landing, fishermen were to state when and where they intended to land, the catch weight on board, and the identity of the fish buyer. All landings were monitored by contracted port validators, who were responsible for determining the weight of halibut and other species off-loaded. The validated halibut weight was then debited from the vessel's IQ. All data were entered into a central computer system within twenty-four hours.

Halibut fishery officers (HFOs) were located throughout the coast to initiate, coordinate, and maintain halibut-related enforcement. The HFOs worked closely with other regional enforcement officers as well as port validators, HAB representatives, and industry participants in an effort to deter infractions, catch violators, and gather information regarding illegal fishing.

Self-enforcement initiatives were incorporated to help the fleet police itself. Representatives discussed enforcement issues at HAB meetings and discussed investigations, activities, charges, convictions, and rumors with the HFOs. The HAB received updated lists of all halibut-related offenses and circulated the information, which resulted in considerable peer pressure being brought to bear on offenders. There was a hot line for fishermen to report illegal fishing activity confidentially.

For court cases involving halibut-related prosecutions, HAB members agreed to provide testimony and evidence on behalf of the Crown as a means of impressing on the courts that the industry, not just the government, was urging compliance. Penalties were serious, including suspension of fishing privileges for flagrant abuse of the program. In 1992, the minister of fisheries and oceans imposed this penalty on a vessel that failed to report its landed catch, and the suspended vessel's IQ for that year was divided among the rest of the fleet.

All costs associated with management, monitoring, and enforcement of the halibut IQ program were recovered from halibut fishermen. A $250 fee plus $0.09 per pound of IQ was collected at issuance of the license. Additional money came from relinquishment of overages. In 1991, the halibut IQ program cost approximately $800,000 to manage, including contracted services, salaries, benefits, overtime, travel, computer programming, vehicle leases, HFO relocations, and other operational expenses and equipment purchases. Fees collected from halibut fishermen totaled $1.2 million. It was estimated that prior to implementation of the IQ program, the DFO spent less than $50,000 annually on management and enforcement for the fishery.

A program evaluation was to quantify and assess economic benefits, product quality, employment, resource effects, safety, enforceability, regional effects, quota-stacking, consultation, and management efficiency and cost-effectiveness. Although a complete evaluation had not taken place by 1998, some quantitative and qualitative results are available.

Program Evaluation

A number of changes occurred while the IQ program was in place. Halibut fishing was spread out over a longer season. Fishermen planned their effort to participate in other fisheries and meet the demands of the lucrative market for fresh halibut. The number of trips per year changed from two, prior to IQs, to as many as fifteen, with an average of three. The number of fishing days varied but increased from a maximum of six before IQs to as many as thirty days of halibut fishing in 1991.

Although the number of crew members employed by halibut vessels in 1991 was estimated to have decreased by 25 percent, the total number of person-hours in the fishery increased in 1991. The number of days fished climbed by about 50 percent. Crew members worked on more than one vessel in the same year.

In the 1990 competitive halibut fishery, the average landed price was Can$2.53 per pound. The average landed price to Canadian halibut fishermen in 1991 was Can$3.04, compared with Can$1.98 for U.S. counterparts continuing in their derby fishery. In 1992, with 70 percent of the Canadian TAC taken, the average price was Can$3.00 per pound, nearly triple the amount U.S. fishermen received for the first 62 percent of their catch. Historically, the respective prices received by U.S. and Canadian halibut fishermen had been comparable.

Under the IQ program, almost all Canadian-caught halibut went directly into the high-quality premium fresh market on a fairly consistent basis throughout the year. More than half of the U.S. catch was of lower quality and frozen, entailing additional handling, freezing, and storage costs.

Considerably less gear was lost as fishermen fished in better weather, set fewer hooks, avoided gear wars, and had time to find and recover lost gear. The result was a reduction in resource wastage. Each year, the IPHC deducted from the exploitable biomass an amount (300,000 pounds prior to setting of the 1992 TAC) for wastage. In 1992, as a result of the IQ program in halibut, the wastage deduction was reduced dramatically. The discarding of less valuable bycatch species, such as rockfish, also decreased as fishermen maximized their fishing operations and the value of all catch brought on board.

Some fishermen chose to use some of their bycatch as bait, further cutting operating costs. If bycatch was not marketable or not wanted by halibut fishermen, they had time to move their fishing operations to grounds where the level of bycatch was lower. Most important, the halibut IQ program resulted in more valuable bycatch data being recorded for fishery scientists, further enhancing stock assessments and resource conservation. This is not to imply that all bycatch was retained or recorded. Bycatch trip limits, market conditions, and a shorter shelf life meant that less than 100 percent of the bycatch was used and recorded.

Perhaps the most significant bycatch problem was in the expanding rockfish longline fishery, in which more than 1,500 active vessels harvested scarce rockfish resources and halibut retention was prohibited, even though halibut was a frequent bycatch. The case was similar in the groundfish trawl fishery, in which juvenile halibut were a common bycatch and the mortality rate was approximately 50 percent.

Traditionally, a majority of the halibut caught annually by Canadian fishermen has been landed in Vancouver, Port Hardy, or Prince Rupert, British Columbia. This remained unchanged. However, Ucluelet, on the western coast of Vancouver Island, became an important port for halibut landings. In addition, small communities remained active in the halibut fishery, with local vessels undertaking day trips, fishing closer to home, and selling their catch to local plants.

Between 1980 and 1990, an average of 18 percent of Canadian halibut caught each year was delivered directly to buyers in the United States. In 1991, only 3 percent of the landed halibut weight was delivered directly to U.S. ports. In 1992, with 70 percent of the TAC taken, approximately 2 percent was delivered directly across the border. More halibut off-loaded in British Columbia meant more off-loading and processing jobs and income staying within provincial and national borders. There were also increased spin-off benefits as fishermen purchased their fuel, ice, and supplies from Canadian suppliers.

There was also a notable change in who was buying the halibut from the fishermen. Traditionally, a majority of the catch was sold to larger processing companies that were well set up, in terms of processing space, freezing capability, and storage capacity, to handle substantial quantities of halibut in a short period of time. Under the competitive fishery, buyers knew months in advance when and how much halibut (usually several million pounds in a few days) would be landed and approximately what share they would get. The larger companies had food market connections with large distributors and retailers who ran sizable advertisements and could sell a lot of halibut in a week or two.

Halibut that was not sold quickly to the fresh market would be frozen and sold throughout the year. Under IQs, landings trickled in throughout the year, often with little prior notice. Large processing companies were not efficiently set up to handle small quantities, nor did they have well-developed market channels to help them move modest amounts into the fresh market.

Smaller companies, however, were well positioned in the market. Prior to IQs, 60–70 percent of Canadian halibut was purchased by large processing companies. In 1991 and 1992, large companies purchased less than 40 percent. Small companies with low overhead, good distribution channels into the fresh market, and efficiency in quickly moving individual vessel catches worked on margins that allowed them to outcompete large, established buyers.

Previously, fishermen did little before fishing to respond to market needs or evaluate other vessels' level of fishing activity and supply of halibut. Many fishermen fished at traditional times because they had done so for years and knew where the fish would be. Fishermen also reacted largely to landed price. When the price was high, a large number of boats would rush out and come back in at the same time. As a result, supply would fluctuate severely, as would prices paid to fishermen. With the IQ program, Canadian halibut fishermen realized the need for a comprehensive marketing strategy, and steps were taken and ideas explored in an effort to smooth out market fluctuations.

In general, enforcement of rules in the fishery was fairly successful, but it was not without problems. Cheating in the halibut fishery did not start with implementation of the IQs. However, IQs did for the first time put a priority on enforcement in the halibut fishery, resulting in the creation of halibut fishery officers. Since very little was expended on monitoring and enforcement before IQs, HFOs realized that a lot of the illegal activity being detected had virtually no association with the IQ program. Infractions such as catching undersized fish, fishing without proper licenses, poaching, and fishing with illegal gear had been occurring for years, but because of the increased enforcement under the IQ program, violators were being caught.

There were also IQ-related violations. Improper notification, off-loading without a port validator present, and catching in excess of a vessel's IQ were some of the more common infractions during the two years of the program. Investigations and charges for more serious violations, such as selling unvalidated catch and providing false documentation, occurred less frequently.

Much of the success of the enforcement plan was due to the high probability of illegal activity being detected. The presence of both port validators and halibut fishery officers throughout the coast, overflights, at-sea boarding by patrol boats, and a fair degree of self-enforcement made it difficult for fishers to cheat without being detected. Experience in the fishery also suggested that peer pressure can be extremely effective in stopping and preventing illegal activity.

As discussed earlier, a common concern in IQ fisheries is high-grading, in which fishers retain only the catch that will garner the highest price. Customarily, large halibut fetched a higher price, and many observers feared that halibut fishermen would retain only large fish to maximize the value of their catch. In practice, however, this did not happen, and for good reasons. First, it turned out that the fresh fish market wanted smaller fish. This worked well in the Canadian fishery, where the exploitable biomass was composed largely of male fish, which are usually smaller than females. Second, Canadian buyers, for the most part, paid across-the-board prices for halibut and did not differentiate by size. Third, operationally, it was very difficult and costly to high-grade.

During the halibut IQ program, as fishermen learned how to better market and distribute their catch and as they became more comfortable with the program, the average value of the halibut quota increased substantially.

Increased market value for halibut licenses also meant that entry into the fishery became more expensive. Moreover, there were large windfall gains to fishermen who held halibut licenses when the IQ program was implemented, an issue that generated considerable controversy. Related issues such as quota concentration, foreign ownership, and recovery of resource rents were under review by the Halibut Advisory Board.

In general, TAC management greatly improved. Every halibut landing was independently observed by a port validator, and the landed weight was entered into a centralized computer system within twenty-four hours. Before these procedures were implemented, the IPHC used weights provided by major processing companies to estimate total catch. The final catch was usually higher when sales slips came in weeks or months later.

It would appear that Canada's Pacific halibut IQ program was successful in meeting various objectives that were set prior to the program, including conserving the resource, improving fishing safety, and increasing economic benefits derived from the halibut fishery.

Notes

1. AS 16.43.010(a). Limited Entry Act of 1973, Alaska Statutes, Chapter 43, Regulation of entry into Alaska commercial fisheries.
2. AS 16.43.010(b).
3. Bruce Twomley, telephone interview, March 15, 1998.
4. AS 16.43.020.
5. AS 16.43.050.
6. AS 16.43.100–120.
7. AS 16.43.140.
8. Twomley, telephone interview.
9. AS 16.43.250(a).
10. Ibid.
11. AS 16.43.250(b)–(c).
12. AS 16.43.225.
13. AS 16.43.230–240.
14. AS 16.43.150.
15. Twomley, telephone interview.
16. AS 16.43.170.
17. Ibid.
18. Twomley, telephone interview.
19. Community Development Quota programs were created under the Magnuson-Stevens Fishery Conservation and Management Act to provide a set-aside of the total allowable catch to rural Alaskan communities.
20. Twomley, telephone interview.
21. Ibid.

22. Ibid.
23. Ibid.
24. Ibid.
25. Ling *(Genypterus blacodes)* is a New Zealand species, not to be confused with lingcod *(Ophiodon elongatus)*, which is mentioned later in this chapter.
26. New Zealand's economy was also suffering from high interest rates and inflation. Subsidies to agriculture, fishing, and other sectors were creating large government budget deficits. A lack of stock assessments for most fisheries prevents precise description of which fisheries were overfished and how badly (Sissenwine and Mace 1992).
27. The revolving fund was never established. Instead, levies paid by industry were used for other government purposes.
28. Orange roughy are extremely long-lived, slow-growing fish that inhabit deep water. Little to nothing was known about their biology well after they had become the target of intensive fishing. When they aggregate for spawning, they are easily captured. As a result, the species is quite vulnerable to overfishing.
29. According to P. Major (1994), the QMS established rights under which the Maori people were able to make legal claims on New Zealand's fishery resources. Absent the system, they would not have been able to make legal claims.
30. The rock lobster and scallop fisheries alone probably added 600 quota owners (Dewees 1996).
31. Resource rent may be defined as "the difference between the market price of a unit of catch and the harvest cost to the marginal producer" (Anderson 1986; Lindner, Campbell, and Bevin 1992).
32. Lindner, Campbell, and Bevin point out that these estimates probably understate the economic value of the fisheries, for several reasons. For instance, conventional accounting procedures commonly seek to underestimate profit.
33. R. O'Boyle, telephone interview, July 2, 1998.
34. Ibid.
35. Ibid.
36. Ibid.
37. Ibid.
38. Ibid.
39. Ibid.
40. Ibid.
41. Ibid.
42. A major obstacle to assigning shares to captains and crew members in

nearly all fisheries is a lack of reliable information (McCay 1994). This lack also prevents accurate evaluation of the effects of ITQs and other programs on employment.

43. The ITQ program did not include the much smaller ocean quahog fishery in Maine, which was managed somewhat independently of the southern fishery.
44. A unique owner may represent several firms.
45. Enforcement was somewhat complicated by a surf clam fishery in state waters of New Jersey, where most surf clams were off-loaded regardless of whether they were caught in federal or state waters. The state fishery was not part of the ITQ program.
46. M. Raizin, telephone interview, March 10, 1998.
47. In adopting the ITQ program in 1990, the Mid-Atlantic Fishery Management Council came under considerable pressure not to require information on such things as prices paid for purchase or lease of quota shares. This has hampered analysis of resource rents and trends in share prices. Indeed, no such analysis has been conducted since Wang and Tang conducted their analysis in 1993.
48. An *otolith* is a calcareous granule found in the inner ear of vertebrates. Over time, layers of calcium carbonate are deposited on the otoliths, allowing scientists to age many species of fish.
49. A fish house is the point at which fishermen unload their catches, the catches are weighed, and the fishermen are paid. The fish house owner records catch details on a form that is forwarded to the relevant government agency.
50. The International Pacific Halibut Commission is the successor to the International Fisheries Commission established in 1923. It was established by the Convention for the Preservation of the Halibut Fishery of the North Pacific Ocean and Bering Sea, Ottawa, March 2, 1953. Protocol, Washington, 1979. (T.I.A.S. 2900).
51. Allocation of shares to boat skippers and crew members was also hampered by a lack of verifiable information, such as audited income tax returns, regarding their participation in the fishery (Pautzke and Oliver 1997).
52. In September 1996, a rule went into effect that allowed an individual holding shares in a larger vessel category to fish those shares from a smaller vessel. P. Smith, letter to Richard Lauber, chairman, North Pacific Fishery Management Council, December 1997 (Internet: <http://www.fakr.noaa.gov/ifq/dec1997.htm>).
53. In the first three years of the IFQ program, uncaught quota shares outweighed the amount by which quota shares were exceeded by individual quota holders. Smith, letter to Richard Lauber.

54. Ibid.
55. Ibid.
56. Ibid.
57. Ibid.
58. Ibid.
59. This case study is adapted from B. R. Turris, "Canada's Pacific Halibut Fishery: A Case Study of an Individual Quota Fishery," pp. 132–151 in *Limiting Access to Marine Fisheries: Keeping the Focus on Conservation,* ed. K. L. Gimbel (Washington, D.C.: Center for Marine Conservation and World Wildlife Fund, 1994).

✦ Chapter 7 ✦

Conclusion: No New Quotas?

The case studies in the previous chapter illustrate that many parts of the world have experienced success in limiting and reducing effort with individual quota programs. Policy makers in the United States, however, remain unsure about the use of this management tool. During the most recent revision of U.S. fisheries law, quota programs went from the center of the debate to off limits.

In a historic exchange on the floor of the House of Representatives in October 1995, members of Congress argued about national fishery management policy for the first time in two decades. Spirited contests engaging members from all over the country arose over a series of floor amendments having national import, each one demanding a roll call vote (*Congressional Record* 1995). Exactly one month short of a year later, the Senate took its turn debating fishery management issues. Nine senators took the podium before an empty chamber on each of two days, either supporting the legislation or offering views and amendments on narrow, regional topics (*Congressional Record* 1996).

The preceding reauthorizations or amendments of the Magnuson-

Stevens Fishery Conservation and Management Act (FCMA),[1] the framework for management of the nearly $4 billion in fishery resources of the country's 200-nautical-mile exclusive economic zone, had been for years the concern of a handful of coastal senators and representatives, "principally a matter of constituent casework" (Iudicello 1992, 60). What took the 1996 reauthorization out of the insiders' circle and into a national policy debate was the convergence of interest on the part of national environmental groups and crises in fishery management that caught the attention of scientists, managers, and national media.

The examination of the rules for use of the public's fishery resources engaged proponents over familiar themes: stewardship of public resources, waste, conflicts of interest, industry's resistance to regulation, cost to the taxpayer of free access to public resources, subsidies for development and extraction of public resources, and, finally, the cost of bailout and restoration after the resource management system failed.

What drew attention to the issue for the 1996 reauthorization was a combination of external events and intentional recruiting and organizing by conservation groups.[2] The groundfish crisis in New England, the high-profile conflict over discards in the pollock fishery in the North Pacific, and the listing of several species of salmon under the Endangered Species Act were among the issues of concern for groups that had not traditionally participated in fishery management issues.

The media helped raise awareness of emerging issues. A cover story in the June 22, 1992, issue of *U.S. News & World Report* titled "The Rape of the Oceans" became a rallying cry for organizers and a target of defensive criticism by the commercial fishing industry. This piece was followed by major stories in such magazines as the *Atlantic,* the *Economist,* and *Mother Jones.*

The four issues that surfaced as defining arguments in the reauthorization were (1) overfishing: what it was, how to stop it, and how to rebuild after it occurred; (2) reduction of bycatch, the incidental capture and discard of nontarget organisms during fishing operations; (3) protection of habitat and reduction of damage from both fishing and nonfishing activities; and (4) whether limited access and quota share programs were viable tools for reducing fishing effort. With the possible exception of habitat protection, each of these issues is related to overcapitalization, or excess fishing capacity.

The FCMA had, since its original passage, allowed for development of limited access programs by eight regional fishery management councils, providing for "a system for limiting access to the fishery in order to achieve optimum yield."[3] These councils and the secretary of commerce were to consider a number of factors, including participation, economics, cultural and social framework of the fishery, and other relevant factors. The law did not, however, clearly state U.S. policy about the nature of the right in a quota share, did not define the terms *individual fishing quota* and *individual transferable quota,* and did not provide a mechanism for initial purchase of quota shares or for collection of rents or fees from shareholders (Mannina 1997).

Dozens of fishermen, academics, managers, environmental advocates, and politicians marched through five years of hearings and submitted mountains of testimony on U.S. fishery management policies. Even though everyone agreed that the United States had "too many boats chasing too few fish," the question of whether the fishery management councils might use quota regimes as a management tool was to become the most contentious issue of all and threatened to unravel the entire reform package in the final months of the 104th Congress. At the heart of national fear and loathing about "privatization of the public's resources" was a decidedly regional squabble over the groundfish of the North Pacific (Gorton 1996; Stevens 1996).

The Alaska fleet is made up primarily of trawlers that deliver to onshore processors, whereas the Seattle-based fleet consists of at-sea catcher-processors, the "factory trawlers." Increased investment, particularly in the latter sector, brought the fleet to the point that industry and agency managers warned it was overcapitalized. In 1989, preparatory to the 1990 reauthorization of the FCMA, Alaska's governor, Steve Cowper, described it this way: "We have a system here . . . that, frankly, leads extensively to over-capitalization, is a quota-driven system that at times puts the gold rush to shame" (Cowper 1989, 73).

The Alaska groundfish complex, which includes the Gulf of Alaska, the Aleutian Islands, and the Bering Sea, is the most abundant of all U.S. fishery resources, contributing nearly 2.2 million tons of catch each year (NMFS 1996b). The species include Pacific halibut, walleye pollock, Pacific cod, sablefish, yellowfin sole, flathead sole and other flatfish, Pacific ocean perch and other rockfish, and Atka mackerel. The total catch of groundfish in this region was more than 1.8 million tons in 1994. U.S. landings of Pacific trawl fish were 4.2 billion pounds, valued at $494.4 million, in 1996 (NMFS 1997a). Most of these species are underutilized or fully utilized according to the National Marine Fisheries Service.

When the fishery began, the fleet was primarily foreign, with a few domestic and foreign joint ventures. By the early 1980s, after implementation of the FCMA and a development program pushed by the state of Alaska, U.S. vessels replaced the foreign fleet. As the fishery grew, allocation among gear types became a more contentious issue at the North Pacific Fishery Management Council (NPFMC). The disputes followed several lines: trawlers versus long-liners, catcher boats and their onshore processor partners versus at-sea catcher-processors, large versus small. Even though the NPFMC continued to set total allowable catches based on the best available science and kept a tight rein on the overall catch, new vessels continued to enter the fishery, making it more difficult to divide the shrinking pie into smaller and smaller pieces among the competing groups. The NPFMC went so far as to ask Congress to place a moratorium on new entrants to the fishery in 1990. The NPFMC finally did, albeit too late, vote a moratorium that one pundit described as "excluding only the unborn from entering the fisheries" (Larkins 1992, 53). Allocation disputes among the various sectors reached a state that approached open warfare. Parties sought and received a specific legislative division of the resource among competing interests. Disputants filed a lawsuit that dragged on for years in an attempt to divide the quota between inshore and offshore processing interests.

On the Atlantic coast of the United States, implementation of quota programs in the surf clam and ocean quahog fishery and the wreckfish fishery was not without debate and controversy. But it was the quota program for halibut and sablefish in Alaska that gained national attention for these programs, coming as it did in the midst of the reauthorization. Moreover, the next quota program under consideration at the NPFMC, for Alaska groundfish, became the whipping post for the entire national debate over limited access, effort reduction, and quota programs. By the time the individual quota (IQ) program for halibut and sablefish was nearing final approval, some sectors of the North Pacific groundfish fleet had already embarked on discussions of what they called rationalization. Subcommittees and task forces of representatives from a variety of fleet sectors discussed how they might design a quota program to reduce fishing effort in the North Pacific.

Limiting of access to U.S. fisheries was discussed at the first hearing to kick off the reauthorization in September 1992, before the Senate Commerce Committee. It came up in every subsequent hearing in both houses, and the House Fisheries Committee devoted an entire hearing to the subject. Proponents of limiting access and allowing fishery management councils to develop individual quota programs cited as benefits an end to derby fishing, increased safety, increased product quality, reduced bycatch, and eventual effort reduction (Mannina 1997, hearing record). Opponents were concerned about privatization of a public resource, windfall profits to quota owners, lack of return to the public in the form of rents or other compensation, ambiguity of conservation benefits, consolidation of quotas by a few large companies, and inequities for crew members (Greer 1995; Mannina 1997). Between the two extremes were those who wanted a clear declaration about the nature of the property right an individual transferable quota (ITQ) would convey and a time-out on new quota programs so that answers to outstanding questions could be found.[4]

Over the course of the reauthorization, numerous bills were introduced, each one containing some provision related to limited access and quota programs. The bill passed by the House, H.R. 39, included an outright prohibition on transferability of quota shares, which IQ proponents saw as a fatal flaw.[5] The bill that was eventually enacted, S. 39, went through numerous drafts on a variety of topics, including the provisions on quota regimes.[6] The changes reflected the growing constituent pressure from Alaskan fishermen on their delegation and pressure on Washington's senators resulting from the economic plight of the Seattle-based catcher-processor fleet.

As introduced, S. 39 defined an individual fishing quota (IFQ) as a "revocable federal authorization to harvest or process a quantity of fish under a unit or quota share that represents a percentage of the total allowable catch of a stock of fish, that may be received or held by a specific person or persons for their exclusive use, and that may be transferred in whole or in part by the holder to another person or persons for their exclusive use."[7] The bill also defined a limited access system as "any system for controlling fishing effort which includes such measures as license limitations, individual transferable quotas, and nontransferable quotas."[8] The bill that was reported about a year and half later dropped the definition of a limited access system and defined an IFQ as a "revocable Federal permit under a limited access system to harvest a quantity of fish that is expressed by a unit or units representing a percentage of the total allowable catch of a fishery that may be received or held for exclusive use by a person."[9] The version finally enacted dropped the word *revocable* and excluded community development quotas from the definition.[10]

The bill as introduced called for promulgation of guidelines for ITQs before the secretary of commerce could approve any fishery management plan containing a quota program, and it set out a list of considerations and requirements that the guidelines were to contain, including conservation, collection of fees, and a reduction in fishing capacity. The original bill also called for amendment of existing plans to bring them into conformity with the guidelines to be promulgated. S. 39 was explicit in defining the nature of the right: "An individual transferable quota does not constitute a property right. Nothing in this section or in any other provision of law shall be construed to limit the authority of the Secretary to terminate or limit such individual transferable quota at any time and without compensation to the holder of such quota."[11] The bill anticipated that quota holders could include vessel owners, fishermen, crew members, other U.S. citizens, and U.S. processors.

By the time the legislation left the Senate Commerce Committee in May 1996, most of this language had disappeared, to be replaced with a prohibition on fishery management councils submitting IFQ plans and a ban on the secretary approving or implementing any plan that included quota programs for four years. The language allowed the councils to revoke existing quota plans without compensation and specifically grandfathered in the potential to amend the existing IFQ programs for surf clams and ocean quahogs, wreckfish, and halibut and sablefish.

Over the summer of 1996, Senator Ted Stevens of Alaska and Senator Slade Gorton of Washington argued back and forth over the quota provisions of the bill. Staff drafts were in a constant state of flux, and the bitter exchanges between the two legislative titans threatened to derail the entire reauthorization. Opponents of ITQs wanted an indefinite moratorium on the use of such measures. Proponents argued for a three-year time-out. Opponents wanted to require a vote by a supermajority of any council to propose an ITQ after the moratorium. Proponents fought the supermajority requirement. Opponents pressed for language that would define *fishing community* in a way that excluded a large city such as Seattle. ITQ proponents wanted no such language. In August 1996, a compromise was struck calling for a four-year moratorium, a detailed and exhaustive study by the National Academy of Sciences, and numerous reports on ITQs to be submitted to Congress in time for study before the next reauthorization. It deleted the requirement for a supermajority council vote on IQ programs and retained a modified version of the language protecting the interests of fishing communities.

Rather than consider the House bill, H.R. 39, Senate sponsors brought their bill, S. 39, to the floor in September. After two days of debate, including lengthy speeches continuing the bitter dispute between Alaska and Washington, the Senate passed the legislation with the compromise provisions. The House took up S. 39 later that month. With reluctance because they had no time to confer on measures they did not like—especially the ITQ provisions, which they termed a "giveaway of a public resource"—they passed the legislation (Studds 1996; Miller 1996).

Table 7.1 compares the quota provisions in the FCMA before reauthorization with the House-passed bill, H.R. 39, and the Senate-passed and enacted bill, S. 39.

The Ocean Studies Board of the National Research Council (NRC) took up the challenge of the ITQ study and empaneled a group of fifteen experts from all over the world with experience in quota programs, science, economics, sociology, business, and anthropology.[12] The NRC undertook an ambitious schedule of public meetings in Anchorage, Seattle, New Orleans, Boston, and Washington, D.C., to solicit views from stakeholders and the public. It is estimated that more

Table 7.1. Comparison of House and Senate FCMA Reauthorization Bills

	FCMA	*H.R. 39, as amended*	*S. 39, as enacted*
Establishment of limited access	Sec. 303(b)(6) allowed limiting of access to U.S. fisheries taking into account present participation; historical fishing practices and dependence on fishery, economics of the fishery; capability of vessels to participate in other fisheries; cultural and social framework; any other relevant considerations.	Amended Sec. 303(b)(6) to authorize establishment of limited access systems if the council and the secretary of commerce take into account need to promote conservation; present participation in the fishery; historical fishing practices and dependence on the fishery; capability of vessels to participate in other fisheries; cultural and social framework and local coastal communities; other relevant considerations.	Amended Sec. 303(b)(6) by changing "limited access" to "system for limiting access to" and by adding "affected fishing communities" to considerations.
Rights; revocability of shares	No provision.	Provided that any quota is considered a grant of permission to engage in activities and can be revoked or limited at any time without compensation; quotas expire no later than seven years after date of issue but can be renewed, reallocated, or reissued at council discretion.	Made clear that IFQs or limited access system authorizations do not create any right or title in fish and that quota shares may be revoked at any time without compensation.
Moratorium on quota programs		Prohibited secretary from approving any quota system plan until completion of review panel and issuance of regulations in accordance with the review.	Created a moratorium until October 1, 2000, on council submission or secretarial approval of any fishery management plan that creates a new IFQ program; allowed amendments to or revocation of existing ITQ programs.
Transferability of quota shares	Allowed transfer.	Prohibited transfer of individual quota shares by sale, transfer, or lease for any new quota system.	Transferability to be considered in study.

Allocation of quotas among users	Councils were given discretion to create allocation systems.	Allowed for allocation of quotas among categories of vessels and provision of portion for entry level, small-boat, and nonqualifying individuals.	Allowed councils to reserve fees to enable purchase of quota shares by small-boat fishermen and first-time participants.
Establishment of fees	Prohibited fee collection, including for quota share programs.	Authorized collection of fees from quota holders by secretary in set amounts to be determined in consideration of the cost of managing the fishery. Did not authorize fee collection for existing quota programs.	Fees to be considered in study.
Restrictions on quota holders	No explicit provision; left to council discretion.	Provided that vessel owners, fishermen, crew members, and U.S. processors could hold quota shares; prohibited individuals not U.S. citizens from holding quota shares.	Limits on ownership to be considered in study.
Review of quota programs	No provision.	Called for review panel to evaluate fishery management plans that establish limited access programs, including quota systems, to examine success of systems in conservation and management; costs of implementation and enforcement; economic effects on local communities; use of leases or auctions in establishing and allocating individual quota shares.	Called for a study by the National Academy of Sciences to evaluate effects of limits on or prohibition of transferability of IFQs; limits on foreign ownership; limits limits on duration of IFQ programs; IFQs for processors; provisions for diversity; adverse social and economic effects; provisions for monitoring and enforcement; threshold criteria for determining fisheries appropriate for IFQ management; mechanisms to ensure fair treatment of owners and crew; potential social and economic costs and benefits to the nation, including fees, auctions, and capital gains revenue; value created for recipients of IFQs; other matters the academy deems appropriate.

than 200 people participated in the process, submitting more than 300 written comments and papers. At the time of this writing, the NRC was embarking on its peer review and internal review process and was expected to release a prepublication version of its results in January 1999.

Although the regional fishery management councils are blocked from submitting quota program proposals to the secretary of commerce, fishermen have been working through and around the moratorium to experiment with other possible measures that create rights-based programs, such as individual vessel bycatch quotas.

At least one commentator has opined that enactment of the Sustainable Fisheries Act of 1996 authorizes IFQs as a form of limited access and that language in the Senate report acknowledges the transferability of IFQs.

In an article in the *Ocean and Coastal Law Journal,* attorney George Mannina, formerly chief counsel to the congressional committee that drafted the Fishery Conservation and Management Act in 1976, says: "Given the historical and often heated debate about the legality of IFQs, the enactment and content of the Sustainable Fisheries Act raises the question: Is there any remaining basis to challenge an IFQ plan as exceeding the authority granted under the Magnuson-Stevens Act?" (Mannina 1997, 11). He concludes that IFQs are a useful conservation tool and a necessary next step in reducing excess capacity in the U.S. fishing fleet (Mannina 1997). Whether the regional fishery management councils will be able to employ this tool is a question for the next reauthorization of the Magnuson-Stevens Fishery Conservation and Management Act.

Notes

1. 16 U.S.C. 1801, Pub. L. 94-265 (as amended) 1976.
2. In a white paper developed for the Surdna Foundation, the Center for Marine Conservation wrote in 1992: "The only way conservation interests are going to be players in the reauthorization of Magnuson is to develop a groundswell of public opinion in favor of conserving America's fishery resources. We have the scientific and policy facts on our side, but in the face of powerful, regional economic interests, they don't hold much sway with Congress. The community needs to

let decision makers know that their constituents include not only harvesters, but consumers, conservationists, concerned citizens who just don't like giving away or wasting the public's resources." The paper laid out the elements of an aggressive campaign to recruit conservation and public interest in fishery management. Upton, H. 1992 Letter from Center for Marine Conservation to Hooper Brooks, Surdna Foundation, conveying report on management of U.S. marine fisheries. February 18, 1992. Unpublished report, 33 pages.

3. 16 U.S.C. 1853(b)(6).
4. R. E. McManus, letter to *Pacific Fishing* regarding ITQs, December 1, 1994.
5. H.R. 39, *An Act to Amend the Magnuson Fishery Conservation and Management Act to Improve Fisheries Management,* October 19, 1995; R. Anderson, "Alaskan Goes Fishing at Northwest's Expense," editorial in *Seattle Times,* April 26, 1995.
6. S. 39, *To Amend the Magnuson Fishery Conservation and Management Act, to Authorize Appropriations, to Provide for Sustainable Fisheries, and for Other Purposes,* January 4, 1995; May 23, 1996 (Report No. 104-276).
7. Sec. 103(19). S.39 (as introduced), *To Amend the Magnuson Fishery Conservation and Management Act, to Authorize Appropriations, to Provide for Sustainable Fisheries, and for Other Purposes,* January 4, 1995. (Introduced by Sen. Stevens for himself and for Sen. Kerry, Sen. Murkowski, Sen. Hollings, Sen. Pressler, Sen. Lott, Sen. Inouye, and Sen. Simpson.)
8. Ibid. Sec. 103(22).
9. S. 39, *To Amend the Magnuson Fishery Conservation and Management Act* (Report No. 104-276). Sec. 103(20).
10. *Sustainable Fisheries Act of 1996,* Pub. L. 104-297, October 11, 1996. Sec. 102(5).
11. S.39 (as introduced). Sec. 111(f)(5).
12. Members of the NRC's Committee to Review Individual Fishing Quotas included Dr. John Annala, New Zealand Ministry of Fisheries; Dr. James Cowan Jr., University of South Alabama; Dr. Keith Criddle, University of Alaska; Dr. Ward Goodenough, University of Pennsylvania; Dr. Susan Hanna, Oregon State University; Dr. Rognvaldur Hannesson, Norwegian School of Economics and Business; Dr. Bonnie McCay, Rutgers University; Dr. Michael Orbach, Duke University Marine Laboratory; Dr. Gisli Palsson, University of Iceland; Ms. Alison Rieser, University of Maine School of Law; Dr. David Sampson,

References

Anderson, L. 1986. *The Economics of Fisheries Management.* Baltimore, Md.: Johns Hopkins University Press.

AQUACULT. 1997. Three diskettes. Rome: Food and Agriculture Organization of the United Nations.

Benton, David. 1997. Telephone interview, Deputy Commissioner for International Fisheries, Alaska Department of Fish and Game, November 19, 1997.

Boyd, R. O., and C. M. Dewees. 1992. "Putting Theory into Practice: Individual Transferable Quotas in New Zealand's Fisheries." *Society and Natural Resources* 5:179–198.

Browning, Robert J. 1974. *Fisheries of the North Pacific: History, Species, Gear, and Processes.* Anchorage: Alaska Northwest Publishing Co.

Buck, S. J. 1998. *The Global Commons: An Introduction.* Washington, D.C.: Island Press.

Burke, W. T. 1994. *The New International Law of Fisheries: UNCLOS 1992 and Beyond.* Oxford, England: Clarendon Press.

Chapman Tripp Sheffield Young. 1997. *Fisheries Act 1996: Overview of the Provisions of the Fisheries Act 1996.* Report prepared for the New Zealand Fishing Industry Board. Auckland, New Zealand: Chapman Tripp Sheffield Young.

Clement & Associates. 1997a. *New Zealand Commercial Fisheries: The Atlas of Area Codes and TACCs, 1997/98.* Tauranga, New Zealand: Clement & Associates.

———. 1997b. *New Zealand Commercial Fisheries: The Guide to the Quota Management System.* Tauranga, New Zealand: Clement & Associates.

CMSER (Commission on Marine Science, Engineering and Resources). 1969. *Our Nation and the Sea: A Plan for National Action.* Washington, D.C.: Government Printing Office.

Congressional Record. 1995. Floor debate on H.R. 39. H10221–47, October 18.

———. 1996a. Floor debate on S. 39, Sustainable Fisheries Act. S10795-10825, September 18.

———. 1996b. Floor debate on S. 39, Sustainable Fisheries Act. S10906-10934, September 19.

Convention for the Preservation of the Halibut Fishery of the North Pacific Ocean and Bering Sea, Ottawa, March 2, 1953. Protocol, Washington, 1979. (T.I.A.S. 2900.)

Cowper, S. 1989. Statement before Committee on Merchant Marine and Fisheries, House of Representatives, on H.R. 2061, August 11, 1989. Serial No. 101-37.

Dewees, C.M. 1996. "Fishing for Profits: New Zealand Fishing Industry Changes for "Pakeha" and Maori with Individual Transferable Quotas." Paper presented at Social Implications of Quota Systems in Fisheries Workshop, May 25–26, Vestman Island, Iceland.

Falloon, R., R. Lattimore, G. Greer, T. Ferguson, and J. Hillman. n.d. *New Zealand Case Study: Trade Effects of Economic Assistance to Fishing.* Canterbury, New Zealand: Lincoln University, MAF Fisheries and Agribusiness and Economics Research Unit.

FAO (Food and Agriculture Organization of the United Nations). 1993. *Marine Fisheries and the Law of the Sea: A Decade of Change.* Special chapter (rev.) in *The State of Food and Agriculture 1992.* FAO Fisheries Circular No. 853. Rome: FAO.

———. 1996. "Approaches to Practical Fisheries Management" by M. K. Kelleher. Round Table on Management and Regulation of Fisheries in the Area of Competence of the Sub-regional Fisheries Commission, July 1–3, Dakar, Senegal.

———. 1997. *Review of the State of World Fishery Resources: Marine Fisheries.* Rome: FAO.

FISHCOMM. 1997. Three diskettes. Rome: Food and Agriculture Organization of the United Nations.

FISHSTAT. 1997. Two diskettes. Rome: Food and Agriculture Organization of the United Nations.

Fordham, S. V. 1996. *New England Groundfish: From Glory to Grief.* Washington, D.C.: Center for Marine Conservation.

Garcia, S. M., and C. Newton. 1997. "Current Situation, Trends, and Prospects in World Capture Fisheries." Pp. 3–27 in *Global Trends: Fisheries Management,* ed. E. K. Pikitch, D. D. Huppert, and M. P. Sissenwine. American Fisheries Society Symposium No. 20. Bethesda, Md.: American Fisheries Society.

Gauvin, J. R. 1994. "The South Atlantic Wreckfish Fishery: A Preliminary Evaluation of the Conservation Effects of a Working ITQ System." Pp. 169–186 in *Limiting Access to Marine Fisheries: Keeping the Focus on Con-*

servation, ed. K. L. Gimbel. Washington, D.C.: Center for Marine Conservation and World Wildlife Fund.

Gauvin, J. R., J. M. Ward, and E. E. Burgess. 1994. "A Description and Evaluation of the Wreckfish Fishery under Individual Transferable Quotas." *Marine Resource Economics* 9 (2): 88–118.

Gimbel, K. L., ed. 1994. *Limiting Access to Marine Fisheries: Keeping the Focus on Conservation.* Washington, D.C.: Center for Marine Conservation and World Wildlife Fund.

Gorton, S. 1996. Remarks on Senate floor during debate on S. 39. *Congressional Record* S10813–14, September 18.

Grainger, R. J. R., and S. M. Garcia. 1996. *Chronicles of Marine Fishery Landings (1950–1994): Trend Analysis and Fisheries Potential.* Fisheries Technical Paper No. 359. Rome: FAO.

———. 1996. "Development Trends and Potential." Pp. 1–12 in *FAO Review of the State of World Fishery Resources: Marine Fisheries.* Rome: Food and Agriculture Organization of the United Nations.

Greer, J. 1995. *The Big Business Takeover of U.S. Fisheries: Privatizing the Oceans through Individual Transferable Quotas (ITQs).* Washington, D.C.: Greenpeace.

Hanna, S. S., C. Folke, and K.-G. Mäler, eds. 1996. *Rights to Nature: Cultural, Economic, Political, and Ecological Principles of Institutions for the Environment.* Washington, D.C.: Island Press.

Hanna, S. S., and S. Jentoft. 1996. "Human Use of the Natural Environment: An Overview of Social and Economic Dimensions." Pp. 35–56 in *Rights to Nature: Cultural, Economic, Political, and Ecological Principles of Institutions for the Environment,* ed. S. S. Hanna, C. Folke, and K.-G. Mäler. Washington, D.C.: Island Press.

Hardin, G. 1968. "Tragedy of the Commons." *Science* 162: 1243–1248.

Hardy, L. F. 1997. *The 1996–1997 Wreckfish Fishery Annual Report.* Beaufort, N.C.: National Marine Fisheries Service, Beaufort Laboratory.

International Pacific Halibut Commission (IPHC). 1997. 1996 Annual Report. Seattle, Wash.

Iudicello, S. 1992. Statement before National Ocean Policy Study, Senate Commerce Committee, hearing on implementation of the Fishery Conservation Amendments of 1990, September 9. S.Hrg. 102-1034 (1993) Washington, D.C.: Government Printing Office.

———. 1996. "Overfishing Lures Legislative Reforms." *Forum for Applied Research and Public Policy* 11 (2): 19–23.

Iudicello, S., S. Burns, and A. Oliver. 1996. "Putting Conservation into the Fishery Conservation and Management Act: The Public Interest in Magnuson Reauthorization." *Tulane Environmental Law Journal* 9 (2): 339–347.

Iudicello, S., and M. Lytle. 1994. "Marine Biodiversity and International Law: Instruments and Institutions That Can Be Used to Conserve Marine Biological Diversity Internationally." *Tulane Environmental Law Journal* 8(1): 123–161.

Jay, Tom, and B. Matsen. 1994. *Reaching Home: Pacific Salmon, Pacific People.* Seattle, Wa.: Alaska Northwest Books.

Johnson, H. M. 1997. *1997 Annual Report on the United States Seafood Industry.* 5th ed. Bellevue, Wash.: H. M. Johnson & Associates.

Keifer, D. R. 1994. "Surf Clam and Ocean Quahog Vessel Allocations, 1977–1992." Pp. 109–117 in *Conserving America's Fisheries: Proceedings of a National Symposium on the Magnuson Act,* ed. R. H. Stroud. National Coalition for Marine Conservation Symposium, March 8–10, 1993, New Orleans, La.

Larkins, H. A. 1992. Statement before National Ocean Policy Study, Senate Commerce Committee, hearing on implementation of the Fishery Conservation Amendments of 1990, September 9. S.Hrg. 102-1034. (1993) Washington, D.C.: Government Printing Office.

Lindner, R. K., H. F. Campbell, and G. F. Bevin. 1992. "Rent Generation during the Transition to a Managed Fishery: The Case of the New Zealand ITQ System." *Marine Resource Economics* 7:229–248.

Love, M. 1996. *Probably More than You Want to Know about the Fishes of the Pacific Coast.* Santa Barbara, Calif.: Really Big Press.

McCay, B. J. 1994. "ITQ Case Study: Atlantic Surf Clam and Ocean Quahog Fishery." Pp. 75–97 in *Limiting Access to Marine Fisheries: Keeping the Focus on Conservation,* ed. K. L. Gimbel. Washington, D.C.: Center for Marine Conservation and World Wildlife Fund.

———. 1996. "Common and Private Concerns." Pp. 111–126 in *Rights to Nature: Cultural, Economic, Political, and Ecological Principles of Institutions for the Environment,* ed. S. S. Hanna, C. Folke, and K.-G. Mäler. Washington, D.C.: Island Press.

McClane, A. J. 1974. *Field Guide to Saltwater Fishes of North America.* New York: Holt, Rinehart and Winston.

McGuinn, A. P. 1998. "Rocking the Boat: Conserving Fisheries and Protecting Jobs." *Worldwatch Paper* No. 142. Washington, D.C.: Worldwatch Institute.

MAFMC (Mid-Atlantic Fishery Management Council). 1996. *Overview of the Surfclam and Ocean Quahog Fisheries and Quota Recommendations for 1997 and 1998.* Dover, Del.: MAFMC.

Major, P. 1994. "Individual Transferable Quotas and Quota Management Systems: A Perspective from the New Zealand Experience." Pp. 98–106 in *Limiting Access to Marine Fisheries: Keeping the Focus on Conservation,* ed.

K. L. Gimbel. Washington, D.C.: Center for Marine Conservation and World Wildlife Fund.
Mannina, G. J. 1997. "Is There a Legal and Conservation Basis for Individual Fishing Quotas?" *Ocean and Coastal Law Journal* (Marine Law Institute, University of Maine School of Law) 3 (1, 2): 5–56.
Milazzo, M. J. 1997. "Reexamining Subsidies in World Fisheries." Unpublished manuscript prepared for the National Marine Fisheries Service, Silver Spring, Md.
———. 1998. *Subsidies in World Fisheries: A Reexamination.* World Bank Technical Paper No. 406, Fisheries Series. Washington, D.C.: World Bank.
Miller, G. 1996. Remarks on House floor during debate on S. 39. *Congressional Record* H11441, September 27.
Monk, G., and G. Hewison. 1994. "A Brief Criticism of the New Zealand Quota Management System." Pp. 107–119 in *Limiting Access to Marine Fisheries: Keeping the Focus on Conservation,* ed. K. L. Gimbel. Washington, D.C.: Center for Marine Conservation and World Wildlife Fund.
NMFS (National Marine Fisheries Service). 1990. *Fisheries of the United States, 1989.* Washington, D.C.: Government Printing Office.
———. 1995. *The IFQ Program, Underway.* Juneau, Alaska: NMFS, Restricted Access Management Division.
———. 1996a. *Fisheries of the United States, 1995.* Washington, D.C.: Government Printing Office.
———. 1996b. *Our Living Oceans: Report on the Status of U.S. Living Marine Resources 1995.* Washington, D.C.: U.S. Department of Commerce and Government Printing Office.
———. 1997a. *Fisheries of the United States, 1996.* Washington, D.C.: U.S. Department of Commerce and Government Printing Office.
———. 1997b. "Reexamining Subsidies in World Fisheries." Unpublished manuscript prepared for the National Marine Fisheries Service, Office of Sustainable Fisheries.
———. 1998. *Report to Congress: Status of Fisheries of the United States.* Silver Spring, Md.: NMFS.
O'Boyle, R., C. Annand, and L. Brander. 1994. "Individual Quotas in the Scotian Shelf Groundfishery off Nova Scotia, Canada." Pp. 152–168 in *Limiting Access to Marine Fisheries: Keeping the Focus on Conservation,* ed. K. L. Gimbel. Washington, D.C.: Center for Marine Conservation and World Wildlife Fund.
OECD (Organization for Economic Cooperation and Development). 1997. *Towards Sustainable Fisheries: Economic Aspects of the Management of Living Marine Resources.* Paris: OECD.
Ostrom, V., and E. Ostrom. 1977. "A Theory for Institutional Analysis of

Common Pool Problems." Pp. 157–172 in *Managing the Commons,* ed. G. Hardin and J. Baden. San Francisco, Calif.: W. H. Freeman.

Ostrom, E., and E. Schlager. 1996. "The Formation of Property Rights." Pp. 127–156 in *Rights to Nature: Cultural, Economic, Political, and Ecological Principles of Institutions for the Environment,* ed. S. S. Hanna, C. Folke, and K.-G. Mäler. Washington, D.C.: Island Press.

Parker, D., and S. Macinko. 1996. *Summary of Executive Summaries: IFQ Research Plan.* Sitka Ak.: Alaska Department of Fish and Game, IFP Research Planning Team.

Pautzke, C. G., and C. W. Oliver. 1997. "Development of the Individual Fishing Quota Program for Sablefish and Halibut Longline Fisheries off Alaska." Paper presented to the National Research Council's Committee to Review Individual Fishing Quotas, September 4, Anchorage, Alaska.

Porter, G. 1997. "The Role of Trade Policies in the Fishing Sector." Background paper for UNEP/WWF Workshop, Fishing Subsidies, Overfishing, and Trade, June 2–3, Geneva, Switzerland.

Raizin, M. 1992. "Individual Transferable Quota Management of the Surf Clam and Ocean Quahog Fishery of the Northwest Atlantic." Paper presented at OECD Workshop on Individual Transferable Quotas, September 18, 1992, Paris. Available from National Marine Fisheries Service, Northeast Regional Office, Gloucester, Mass.

Roodman, D. M. 1996. *Paying the Piper: Subsidies, Politics, and the Environment.* Worldwatch Paper No. 133. Washington, D.C.: Worldwatch Institute.

Rosenberg, A. A. 1994. "Background on U.S. Fisheries Management: Status and New Directions." Pp. 27–41 in *Limiting Access to Marine Fisheries: Keeping the Focus on Conservation,* ed. K. L. Gimbel. Washington, D.C.: Center for Marine Conservation and World Wildlife Fund.

SAFMC (South Atlantic Fishery Management Council). 1990. *Snapper Grouper Assessment Group: Wreckfish Report.* Charleston, S.C.: SAFMC.

———. 1993. *Snapper Grouper Assessment Group: Wreckfish Report.* Charleston, S.C.: SAFMC.

———. 1997. *Snapper Grouper Assessment Group: Wreckfish Report.* Charleston, S.C.: SAFMC.

Schrank, W. E. 1997. "The Newfoundland Fishery: Past, Present, and Future." Pp. 35–70 in *Subsidies and Depletion of World Fisheries.* Godalming, England: World Wide Fund for Nature.

Sissenwine, M. P. 1998. "Is There Something Wrong with the Medicine, or Does It Just Taste Bad?" Paper presented at conference, Year of the Oceans, March 3–4, John Heinz III Center for Science, Economics and the Environment, Washington, D.C.

Sissenwine, M. P., and P. M. Mace. 1992. "ITQs in New Zealand: The Era of Fixed Quota in Perpetuity." *Fishery Bulletin* 90:147–160.

Sissenwine, M. P., and R. A. Rosenberg. 1993. "Marine Fisheries at a Critical Juncture." *Fisheries* 18 (10): 6–11.

Stevens, T. 1996. Remarks on Senate floor during debate on S. 39. *Congressional Record* S10810, September 18.

Studds, G. 1996. Remarks on House floor during debate on S. 39. *Congressional Record* H11440, September 27.

Turris, B. R. 1994. "Canada's Pacific Halibut Fishery: A Case Study of an Individual Quota Fishery." Pp. 132–151 in *Limiting Access to Marine Fisheries: Keeping the Focus on Conservation,* ed. K. L. Gimbel. Washington, D.C.: Center for Marine Conservation and World Wildlife Fund.

Twomley, B. 1994. "License Limitation in Alaskan Salmon Fisheries." Pp. 59–66 in *Limiting Access to Marine Fisheries: Keeping the Focus on Conservation,* ed. K. L. Gimbel. Washington, D.C.: Center for Marine Conservation and World Wildlife Fund.

Upton, H. F., P. Hoar, and M. Upton. 1992. *The Gulf of Mexico Shrimp Fishery: Profile of a Valuable National Resource.* Washington, D.C.: Center for Marine Conservation.

Wang, S. D., and V. H. Tang. 1993. *The Performance of U.S. Atlantic Surf Clam and Ocean Quahog Fisheries under Limited Entry and Individual Transferable Quota Systems.* Gloucester, Mass.: National Marine Fisheries Service, Northeast Regional Office.

Warming, J. 1983. "On Rent of Fishing Grounds." *History of Political Economy* 15(3): 391–396. Peder Anderson translation of Jens Warming's 1929 model.

Weber, M. L. 1997a. "Effects of Japanese Government Subsidies of Distant Water Tuna Fleets." Pp. 119–136 in *Subsidies and Depletion of World Fisheries: Case Studies.* Washington, D.C.: World Wildlife Fund.

Weber, M. L., and J. A. Gradwohl. 1995. *The Wealth of Oceans: Environment and Development on Our Ocean Planet.* New York: Norton.

Welch, L. 1997. "It's a Halibut Heyday." *National Fisherman* (June): 16–18.

Wieland, R. 1992. *Why People Catch Too Many Fish: A Discussion of Fishing and Economic Incentives.* Washington, D.C.: Center for Marine Conservation.

Wildman, M. 1993. *World Fishing Fleets: An Analysis of Distant-Water Fleet Operations.* Vol. 3, *Asia.* Silver Spring, Md.: National Oceanic and Atmospheric Administration, Office of International Affairs.

Wilen, J. E., and F. Homans. 1997. "Unraveling Rent Losses in Modern Fisheries: Production, Market, or Regulatory Inefficiencies?" Pp. 256–263 in *Global Trends: Fisheries Management,* ed. E. L. Pikitch, D. D. Huppert, and M. P. Sissenwine. American Fisheries Society Symposium No. 20. Bethesda, Md.: American Fisheries Society.

Index